| DATE DUE | | | |
|---|---|---|---|
| | | | |
| | | | |
| | | | |
| | | | |
| | | | |
| | | | |
| | | | |
| | | | |
| | | | |
| | | | |
| | | | |
| | | | |

# THE BIOCHEMISTRY OF ANIMAL FOSSILS

# THE BIOCHEMISTRY OF ANIMAL FOSSILS

BY

RALPH W. G. WYCKOFF

*Professor of Physics and Microbiology, The University of Arizona, Tucson, Arizona*

BRISTOL: SCIENTECHNICA (Publishers) LTD.

**The Williams & Wilkins Company**

**Baltimore**

*Distribution by Sole Agents:*
*United States of America: The Williams & Wilkins Company, Baltimore*
*Canada: The Macmillan Company of Canada Ltd., Toronto*

ISBN 0 85608 004 7

PRINTED IN GREAT BRITAIN BY JOHN WRIGHT AND SONS LTD.,
AT THE STONEBRIDGE PRESS, BRISTOL BS4 5NU

# PREFACE

THE general objectives and contents of this monograph are stated in its first, introductory, chapter and need not be repeated. It deals only with the still limited number of studies of single fossil specimens, as contrasted with fossiliferous masses, that have been made worth while by the development of chromatographic methods of analysis. Though still largely exploratory and fragmentary, they clearly indicate how knowledge of the internal structure and composition of such specimens can add to our understanding of evolution, and perhaps of the earliest forms of life and the way they came into being.

Many of the data presented here, collected in the author's laboratory, are being published for the first time. He is indebted to various collaborators in their collection and especially to Franklin D. Davidson for his help in the development of techniques and the preparation of illustrative material. Experimental work has been supported by Grants No. 5 T01 DE00126 and 5 R01 DE01919 from the National Institutes of Health and No. GB-17516 from the National Science Foundation.

*Tucson, Arizona.* R. W. G. W.
*January*, 1972.

# CONTENTS

# THE BIOCHEMISTRY OF ANIMAL FOSSILS

## *CHAPTER I*

## INTRODUCTION

Every thinking man today must be acutely aware that the growth of natural science has changed the meaning life can have for him. This meaning has ceased to be that which shaped our inherited culture and guided the daily lives of our forebears, for we can no longer look upon ourselves as special creations privileged to enjoy Nature's bounties. Instead, our lives must be built around a realization that as men we are but one stage in an evolutionary process which has been going on for hundreds of millions of years and which began with inconceivably simple forms of living matter. To understand ourselves and gain some insight into what mankind's future can be, we need to learn everything possible about these primordial origins and the unbroken series of steps that lie between them and ourselves.

Most of what we have learned about this evolutionary process has come through the detailed study and comparison of remains of plant and animal life which, embedded in rocks, form part of the geological record of the Earth's earlier history. The science of palaeontology has grown out of a precise description of the external appearance of these fossils and through reconstructing the skeletal remains of which they formed part. This work has now been carried to the point where detailed pictures can be given of the successions of organisms that have flourished and then died off in the past or have given rise to the innumerable forms of existing life. In some of these continuing streams of life there has been progression from a simple to a more complex state of organization; in others, evolutionary development long ago became arrested through an environmental adaptation so perfect that the animal living today is scarcely to be distinguished from its ancestors of hundreds of millions of years ago.

Though palaeontology has thus given a remarkably detailed insight into the chain of life, much more must be learned before we can adequately understand either how life originated in the first place or what factors determine its continuing evolution. This additional knowledge will come in part through the biology that intimately studies life as it functions today. At the same time biochemistry has now grown until the composition of living matter can be compared in new ways with the remains of earlier life, and meaningful experiments can begin to reproduce in the laboratory the conditions under which the first living forms appeared.

To the palaeontologist engaged in his central task of reconstructing the macroscopic form of an ancient animal from its fossil parts, it is of secondary importance whether they are the original bones or are casts produced through gradual replacement by extraneous minerals. Partly for this reason and partly because adequate techniques did not yet exist, relatively little attention has in the past been given to the internal structure and composition of the fossils themselves. Teeth and shells have been extensively examined under the polarizing optical microscope, but the thoroughgoing investigations that now are possible have had to await the invention of the vastly more powerful electron microscope and of new, more sensitive methods of chemical analysis based on chromatography.

Various things are to be learned through such investigations. We know that evolution has been accompanied by, and in a sense expressed through, an increasing complexity in the organization of plants and animals. Have their parts also become more complicated either in internal structure or in chemical composition? Examination at very high magnifications can show if the present internal structure of fossils is that which prevailed in the living animal; and by comparing this intimate structure with that of present-day animals we can ascertain in what ways microstructure may have altered during the course of evolution.

The question of chemical evolution must be approached in a less direct fashion. It seems certain that life began in an environment very different from that which now prevails; the atmosphere was reducing rather than oxidizing and temperatures were higher. This atmosphere and seas, shallower and less saline than now, were presumably sites of vigorous chemical activity. Recent laboratory experimentation has shown that amino-acids and other simple constituents of living matter form under such conditions and it is reasonable to assume that compounds produced in this abiotic way became part of the first living organisms. We are, however, still ignorant of how such compounds may have combined to form the proteins and other complicated substances whose organized interactions characterize the living state. Were these essential ingredients of the earliest life as complex as those encountered today, or did life proceed at first through a comparatively simple series of biochemical reactions? This is the question raised by the possibility of a chemical, paralleling the established morphological, evolution of living forms.

The chemical residues of earlier plant life have long been the subject of intensive investigation largely because they constitute the world's most extensively exploited source of energy. Fossilized fragments—leaves, stems, tree trunks, and pollen—are found in coal but the chemical changes that have produced it are so profound that the many compounds isolated from it, and from crude oil, are far removed from those of which the living plants were composed. Many animal fossils, on the contrary, are sufficiently well preserved to ensure that the structures and substances found within them originated in the creature from which they were derived. This, together with our new ability to analyse in detail the retained products of initial vital activity, gives the study of such fossils its great promise. They are the main subject of this monograph. The work carried out on them is still exploratory but the results, though necessarily preliminary, are sufficient to indicate what can be learned.

Succeeding chapters will consider first (in CHAPTER II) what modern techniques can teach about the fine structure of fossils and, in particular, the state of their preservation. It does not aim to describe in detail the complicated structures that these techniques have already revealed in well-preserved specimens, or the growing number of investigations utilizing one or another of them to help find the place of certain early forms within the chain of evolution. Structural considerations enter only as they supply the essential basis for meaningful analyses of the products of vital activity preserved within individual fossils. CHAPTER III is a brief discussion of the composition of both the inorganic and organic components of animal fossils and an outline of methods useful in analysing these components and dealing with the problems they bring up. The focus of interest is in CHAPTER IV. If there has been chemical evolution, it probably occurred in the proteins which are the essential, characteristic ingredients of all living matter. This makes any retained proteins and their immediate degradation products of special interest. Their study and the results thus far obtained are more exhaustively treated in this chapter. Other universal components of living matter, notably fats and polysaccharides, occur in some specimens and are briefly discussed in CHAPTER V. CHAPTER VI is a short summary of this preliminary work, emphasizing some of the most promising directions in which it may be expected to develop.

It would be of the greatest value to find out something about the nucleic acids of early forms of life. Their chemical reactivity, however, makes a prolonged survival unlikely, and as yet no serious search for them appears to have been made. There have been numerous attempts in the past to deal with the organic matter in fossils, but the development of quantitative chromatography has left them little more than a historical interest. References to them have not been included in the Bibliography that appears as an appendix to the book, but for those who wish it several of the reviews cited are avenues to this older literature.

## *CHAPTER II*

# THE MICROSTRUCTURE OF FOSSILS

DETERMINATIONS of structure are attempts to visualize in one way or another the component parts of an object and their interrelations. All studies of microstructure must therefore be exercises in microscopy and are dependent for their advance on the development of more powerful and versatile instruments for this purpose.

When examining a fossil, the first thing to be ascertained is the extent to which the original structure has been preserved. In order properly to interpret observations we now make, it is vital to know if what is seen is detail that was present when the animal was alive, or an artefact developed either on death or while the animal remains lay embedded in rock. Even when the original material of bone or shell has been retained, it may have recrystallized with time, or the structural details may have become distorted under the pressures and elevated temperatures to which a fossil and its enclosing matrix may have been subjected.

As a matter of fact all degrees of preservation are found. Many fossils, even at the highest magnifications of the electron microscope, exhibit an internal structure indistinguishable from that to be seen in living material. At the other extreme, some fossils are no more than casts of a living bone or tooth; all the original tissue, hard and soft alike, has been dissolved away and replaced by entirely extraneous minerals. At times this replacement reproduces with remarkable fidelity details that were present in life, but this is rare; more often it is a waste of time to study a heavily altered fossil for evidence of what its internal structure and contents may have been.

Gross evidence of alteration can be obtained by examining a fossil at optical microscopic magnifications, either after first etching a polished surface or as a thin section viewed under polarized light. In this laboratory* such an examination is preliminary to all further work. Specimens are made by embedding a fossil fragment in a thermo-setting plastic for ease in handling, and cutting slices approximately a millimetre thick from it with a diamond saw. For photography at low magnifications one face is ground, polished, and etched with either hydrochloric or formic acid. The strength of acid and the duration of etch will depend on the specimen. Sometimes this surface, obliquely illuminated to bring out its details, is photographed directly. More can, however, usually be seen after evaporation on it of a thin film of gold.

Thin sections for microscopical examination are made by grinding a slice to the desired thinness either by hand or with one of the more or less automatic grinders employed by petrographers. Hand grinding can be done on a glass plate with powdered abrasive. The slice must be rigidly held to obtain a final section with flat and parallel sides. Numerous procedures have been described for doing this. One which gives sections of a uniform thickness of 75 microns uses orienting blocks made by Buehler Ltd. When, as is often the case, still thinner sections are wanted, they are made by a further grinding of these petrographic sections with the help of a modified piston-cylinder such as that employed when preparing samples for reflection microscopy. This tool is made by giving the combination a well-polished common face that is very exactly normal to its axis. The section being ground is mounted on a slide that fits closely within the cylinder and is held in position by pressure on the piston. In this way extremely thin sections only a few microns thick can be successfully prepared.

These sections are sometimes most profitably examined under the polarizing microscope, with or without insertion of a quarter-wave plate to introduce colour differences. With many other preparations more detail can be seen, especially at the higher magnifications, by taking advantage of the three-dimensional impression created by the slightly out-of-phase beams of Nomarski interference illumination.

The types of structural information to be gained from photographs of etched surfaces and micrographs of sections are different for fossil bones, teeth, and shells. Though bones may differ in

* Throughout this monograph the term 'this laboratory' refers to that of the writer in the Department of Physics at the University of Arizona, Tucson, Arizona.

porosity and amount of organic matter, and consequently in strength, fresh and well-preserved specimens present very much the same features irrespective of type of animal. As *Fig.* 1 illustrates, sections exhibit under the polarizing microscope strongly marked patterns of double refraction due to the parallel alinement of groups of submicroscopic apatite crystals. When the changes

*Fig.* 1.—V63210. A thin section of Pleistocene horse bone viewed under crossed Nicols in the polarizing microscope. In the living state the elliptical holes of the osteons, here filled with calcite crystals of various sizes, contained cells. Parallel alinement of submicroscopic birefringent crystallites is responsible for the concentric patterns of light and dark of the apatitic framework. (×90.)

that have occurred during fossilization have not seriously disturbed this internal arrangement, the double refraction is indistinguishable from that seen in sections of fresh bone. The same detail is apparent in the Nomarski image of *Fig.* 2. In the living state organic matter, largely as collagen, is distributed throughout this meshwork of apatite crystallites, and the osteons and lacunae contain living cells with their numerous filamentous processes. In most fossils the cells have disintegrated and at least some of the collagen has disappeared. In both these figures the masses of secondary calcite which have invaded the bone and formed in the voids can readily be identified in the prepared sections by

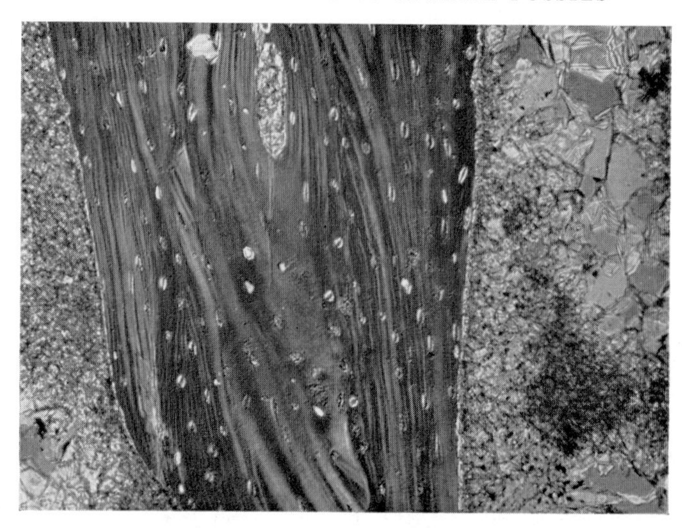

*Fig.* 2.—V63210. A thin section of fossil horse bone cut at right angles to that of *Fig.* 1 and viewed under Nomarski illumination. Cellular channels, now filled with calcite, are on either side of the apatitic wall which is perforated by many lacunae. In life these were occupied by processes of the cells that filled the channels. (×225.)

*Fig.* 3.—V67138. The polished and etched section through a Jurassic dinosaur bone photographed under oblique illumination. The corrugated surface produced by the etching is due to the same preferred crystallite orientation responsible for the birefringence of *Fig.* 1. The lighter-coloured mineral filling the cellular spaces in this spongy bone is most apparent to the right of centre. (×45.)

standard petrographic procedures. The substructure of the apatite responsible for its birefringence and shown in the preceding two figures is developed by etching and can be seen even at low magnification (*Fig.* 3). In this photograph, too, the secondary minerals that fill the spaces initially occupied by cells are easily recognized. Damage by pressure or shearing is manifested by such cracks in the apatitic framework as those to be seen in *Fig.* 1. When a fossil has been permeated by acidic solutions, the apatite may be partially dissolved and replaced by masses of newly formed, disoriented crystals whose birefringence does not conform to that evident in the figures. Occasionally fossil bones are encountered with abnormalities that suggest damage or disease during the life of the animal. Many years ago this palaeopathology was a subject of active, though limited, concern (Moodie, 1923). New microscopic techniques such as those being discussed here make it once more a most promising field for future research (Ascenzi, 1955; Tasnádi-Kubacska, 1962; Jarcho, 1966; Erben, 1969b). They can bring to light abnormalities that could not earlier have been seen and remove any difficulties there may have been in distinguishing damage incurred during life from that arising after fossilization had begun.

The teeth of most vertebrates, irrespective of geological age, are composed of the same structural elements: an outer *enamel* layer that is almost pure apatite, a thicker layer of *dentine* which is a calcified mass of collagen situated between the enamel and the innermost *pulp* that houses the living cells fed by blood and nerves. In addition, mammalian teeth possess a calcified mass of cells (the *cementum*) interposed between the root and jawbone; in browsing animals this layer of cementum may be greatly thickened and interleaved between folds of enamel and dentine. The microcrystals of apatite are sufficiently alined to make enamel doubly refracting under the polarizing microscope, but elsewhere in a tooth their disorder is such that the structure is better seen with other types of illumination. The importance of the shape of teeth and their arrangement in the jaw as a means of classifying extinct animals has led to older studies under polarized light recorded in Scott and Symons (1964) and Peyer (1968). These observations have not, however, been greatly expanded to take advantage of the more recent developments in microscopy.

Mature enamel contains so little organic matter that well-preserved fossil enamel cannot be distinguished from fresh. It consists of compact groups of submicroscopic crystals alined to form rods (*Figs.* 4 and 5), themselves loosely or tightly arranged in bundles whose interlocking gives enamel its strength and

*Fig.* 4.—V633. Enamel rods in a thin section through the enamel of a Miocene desmostylus tooth. Bundles of rods roughly parallel to one another produce the broad light and dark bands of this photograph. Nomarski illumination. (×225.)

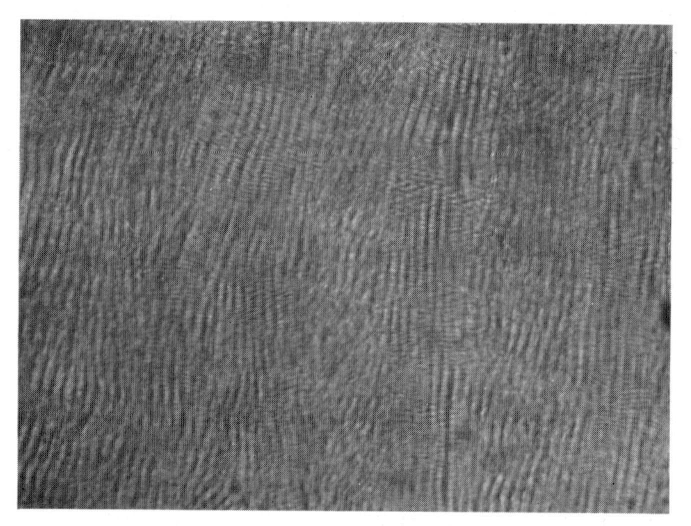

*Fig.* 5.—V64145. Enamel rods in a thin section of the tooth of a Pleistocene sabre-tooth tiger showing their fine, transversely segmented character. Nomarski illumination. (×360.)

durability. These rods are clearly seen in sections of appropriate thickness and their dimensions, details of internal structure (*Fig.* 5), and modes of packing (*Figs.* 6 and 7) can there be studied.

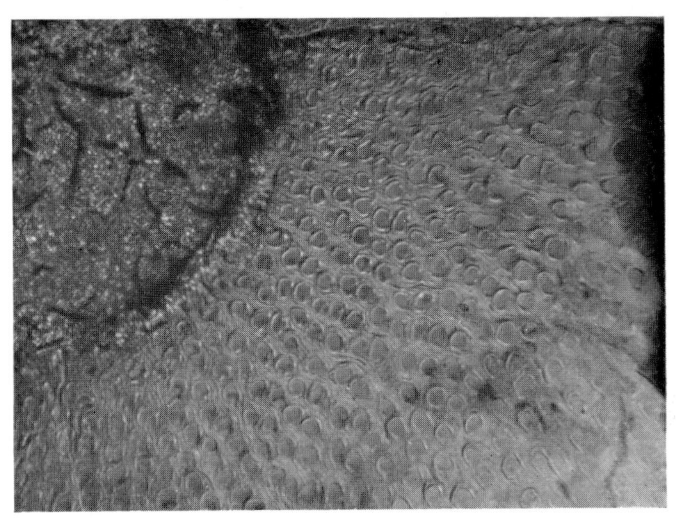

*Fig.* 6.—V697. A cross-section through the cylindrical bundles in the enamel of a Palaeocene taeniolabid tooth. The individual units that are packed together within these bundles are barely discernible in the upper part of the photograph. Nomarski illumination. (×360.)

*Fig.* 7.—V63184. Ends of the enamel rods exposed in a section of a Pleistocene mastodon molar showing their tight, orderly packing. (×675.)

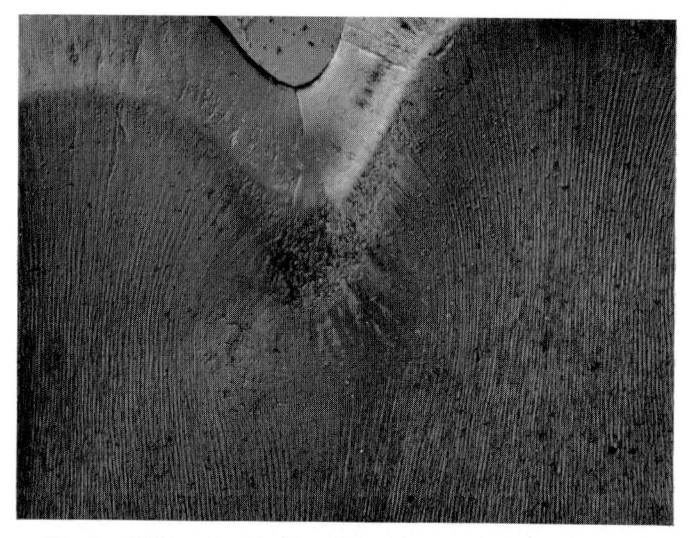

*Fig.* 8.—V6860. A thin section through the serrated tooth of a Cretaceous dinosaur showing dentine with its innumerable tubules capped by a thin, transparent layer of enamel. Nomarski illumination. (×225.)

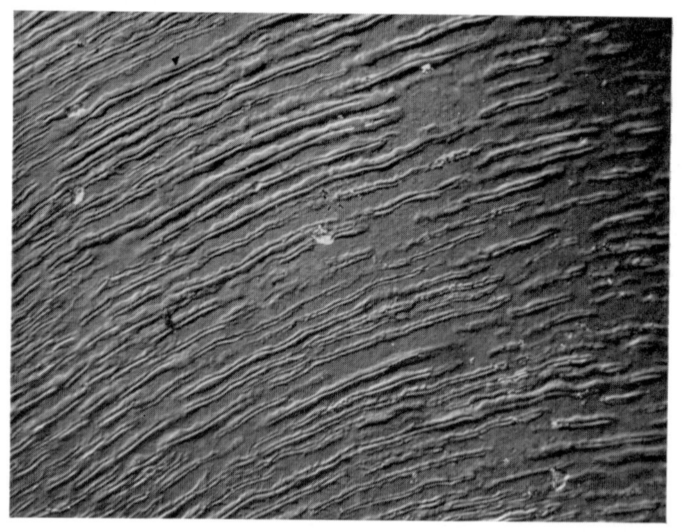

*Fig.* 9.—V63217. An especially thin section through the dentine of a Pleistocene horse molar. The more refractive walls and the hollow cores of the individual tubules are apparent. Nomarski illumination. (×225.)

Any chemical attack upon them and disruptions in packing during fossilization cannot be missed. Dentine always appears as a system of minute tubules radiating outwards from the pulp through a calcified matrix (*Fig.* 8). In living teeth these tubules (*Fig.* 9), embedded in a less dense matrix, contain cellular processes. In fossils the organic filling of the tubules has largely disappeared and has frequently been replaced by extraneous minerals. In a

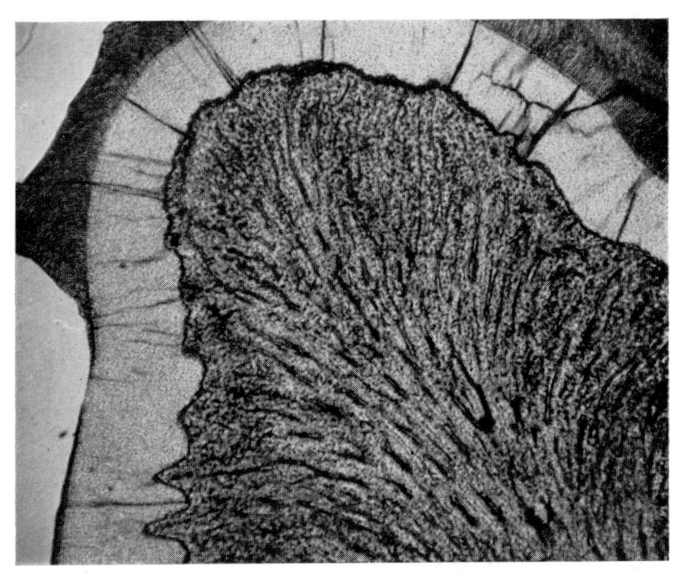

*Fig.* 10.—V6351. A section through a Pleistocene horse tooth. Cementum with its extensive system of cellular processes occupies the major portion of the photograph. Above it is a clear band of enamel surmounted by a strip of dentine with tubules barely visible in the upper right-hand corner. Nomarski illumination. (×37·5.)

few fossil teeth one sees in the pulp cavity what appear to be the remains of its cellular contents; more often it is either empty or filled with secondary minerals. Wherever cementum is an important component it is easily recognized by the characteristic texture given it by its many cells. It occupies the central part of the micrograph of *Fig.* 10, made of a thin section through the tooth of a Pleistocene horse. In *Fig.* 11 is to be seen the region where the cementum of a Miocene sloth tooth meets dentine.

Hard tissues composed mainly of calcium carbonate, and their fossil remains, are more widespread than the phosphatic bones and teeth. Two crystalline forms of the carbonate, calcite and aragonite, occur as minerals and are found in fossils. The egg-shells of birds and of most reptiles that have hardened shells are of calcite. Some of the latter are aragonitic and this mineral is also frequently encountered in the shells of invertebrates (Stehli, 1956; Hall and Kennedy, 1967). Though a few primitive invertebrates produce apatite and the exoskeletons of diatoms and

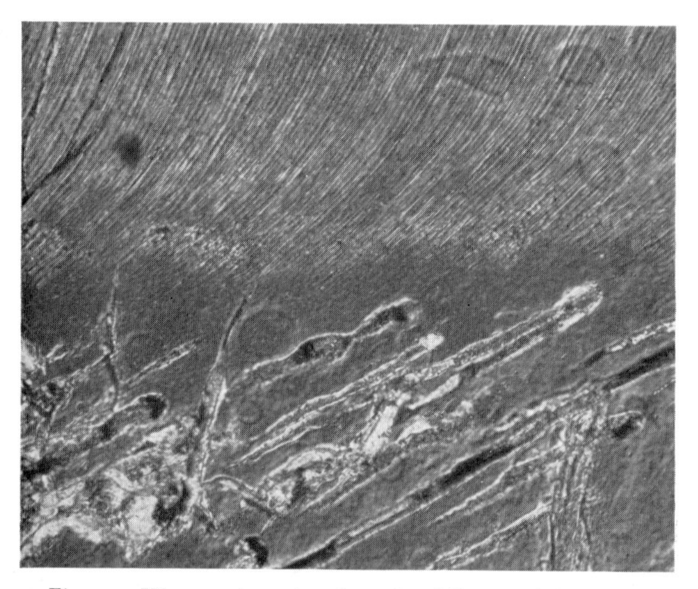

*Fig.* 11.—V64151. A section through a Miocene sloth tooth. At this higher magnification the fossilized cellular processes that permeate cementum (*below*) are better seen than in the preceding figure. Nomarski illumination. (×225.)

certain other micro-invertebrates are siliceous, the hard tissues of all other invertebrates are either of organic chitin, sometimes strengthened with mineral, or of nearly pure calcium carbonate.

The crystalline egg-shells of birds enclose proteinaceous membranes. In the thin shells of most birds (Terepka, 1963) arrangement of the crystals of calcite is generally rather poor, but in the thick shells of large ratite birds and their fossil forebears it is more orderly. The central portion of a cassowary egg-shell, for instance, is a finely crystalline mass containing much granular and membranous organic matter, capped on either side by layers of large calcite crystals (*Fig.* 12). In ostrich shell, on the contrary,

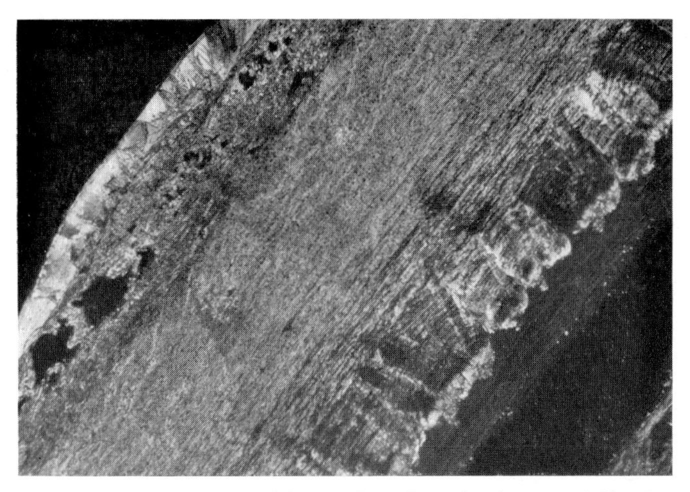

*Fig.* 12.—V6896. A thin section through the egg-shell of a cassowary seen under crossed Nicols. The coarsely crystalline mammillae of the inner layer of the shell are to the right; the thinner but equally coarse pigmented layer decorating the shell is to the left. Much organic matter appearing black in the figure is dispersed throughout the central region. (× 37·5.)

*Fig.* 13.—V6897. A section similar to the preceding through the shell of an ostrich egg seen under crossed Nicols. Single crystals extend without interruption from the mammillae on the right across the entire shell. The many organic membranes embedded in the shell appear as thin lines, parallel to one another and to the borders of the shell. (× 37·5.)

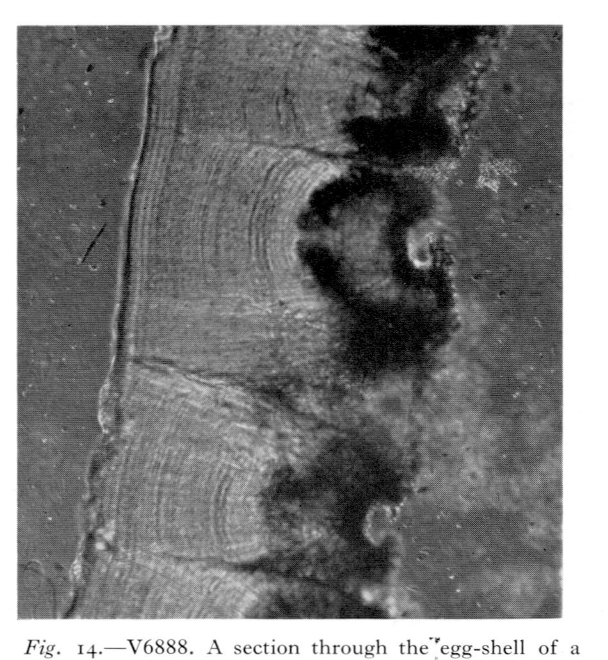

*Fig.* 14.—V6888. A section through the egg-shell of a desert tortoise. As in the ostrich egg-shell large single crystals reach from edge to edge but in this case they are of aragonite. The mammillary region is partially obscured by black organic matter. Nomarski illumination. (×225.)

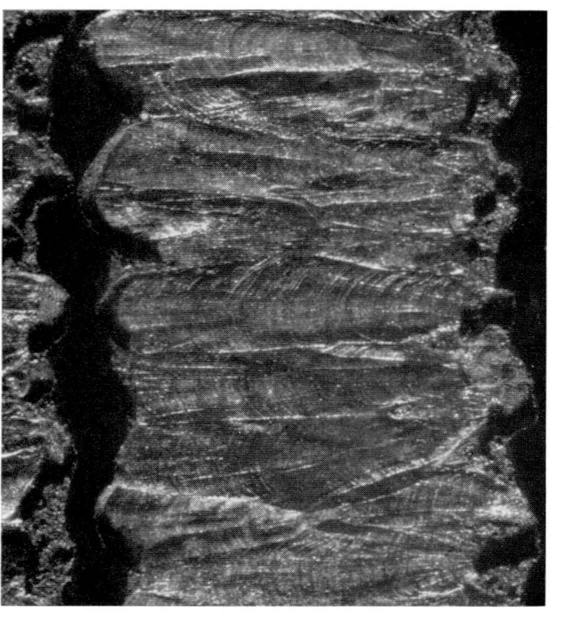

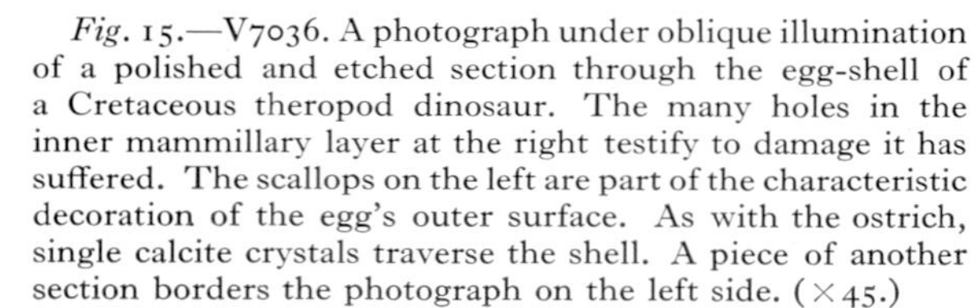

*Fig.* 15.—V7036. A photograph under oblique illumination of a polished and etched section through the egg-shell of a Cretaceous theropod dinosaur. The many holes in the inner mammillary layer at the right testify to damage it has suffered. The scallops on the left are part of the characteristic decoration of the egg's outer surface. As with the ostrich, single calcite crystals traverse the shell. A piece of another section borders the photograph on the left side. (×45.)

the large crystals extend across the entire shell, thus traversing without interruption innumerable membranes (*Fig.* 13). The hardened shells of modern reptilian eggs are equally diverse in structure. Some consist of disordered masses of crystals deposited on a thick organic membrane; in others, such as the desert tortoise, the shell resembles that of the ostrich in being an ordered assemblage of large crystals (*Fig.* 14), but for this and other turtles the

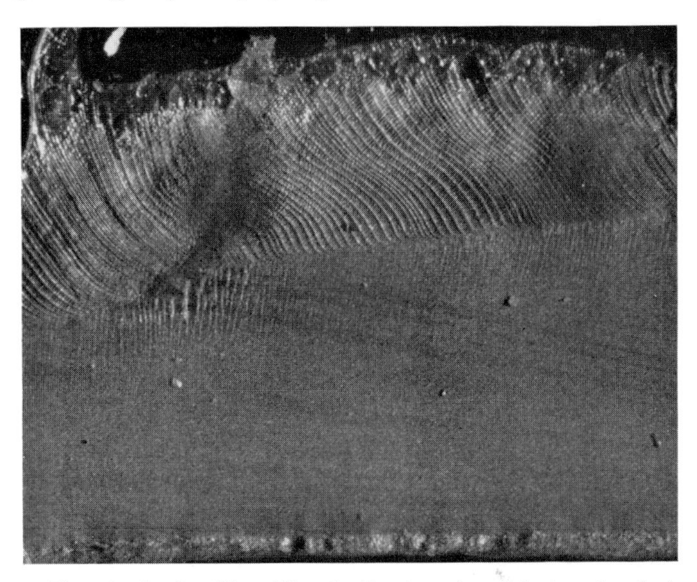

*Fig.* 16.—17080. The obliquely illuminated, polished, and etched surface of a section through the aragonitic shell of *Chione gnidia* showing (*above*) its outer composite prismatic and (*below*) its inner homogeneous structure. (×45.)

mineral is aragonite. The shells of dinosaur eggs, too, are composed of large single crystals which like those of the ostrich are of calcite (*Fig.* 15).

The thick carbonate shells of marine invertebrates often have complex structures. Earlier examination with the polarizing microscope of both Modern and fossil shells has led to their classification in terms of the way their crystals are arranged. More than one type may prevail in a single shell and it would appear that membranes of somewhat different compositions may be associated with each (e.g., Ranson, 1952; Wada, 1966). The principal types of arrangement, recognizable in both fresh and fossil specimens, have been designated as homogeneous,

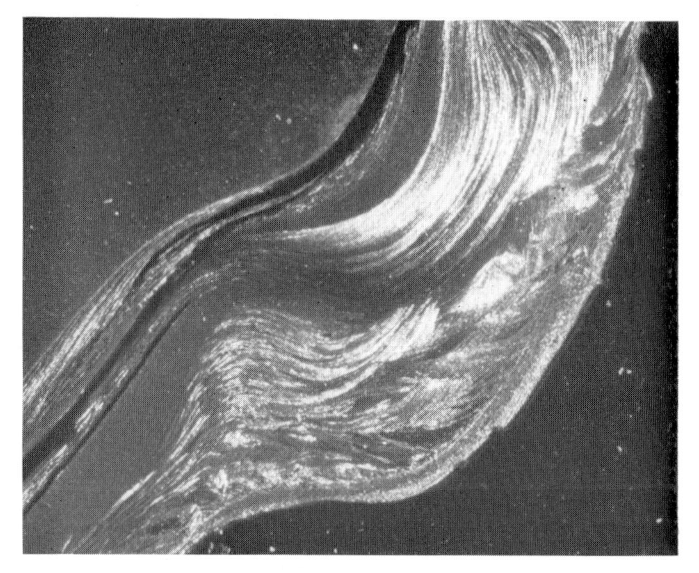

*Fig.* 17.—I7031. The obliquely illuminated, polished, and etched section through the calcitic shell of *Chlamys eboreus* showing its characteristically foliated structure. (×45.)

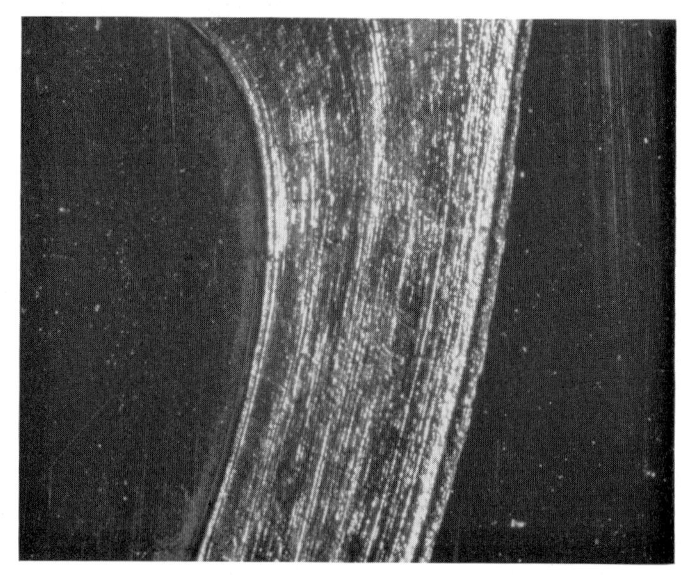

*Fig.* 18.—I711. An obliquely illuminated, polished, and etched section through the nacreous, aragonitic shell of *Pinctada* sp. (×45.)

prismatic, foliated, crossed-lamellar, and nacreous (Bøggild, 1930). Either calcite or aragonite may be present in one of these types. Among the molluscs the homogeneous part of their shells is in some cases calcite, in others aragonite; in other invertebrates the homogeneous type is of calcite. The structure of a shell is often better recognized in etched section than under polarized light. In such a section the homogeneous type is characterized by its apparent lack of fine structure (*Fig.* 16). An example of the foliated, which is always composed of calcite, is shown in *Fig.* 17. The minute, parallel crystals that give the nacreous portions of a shell (*Fig.* 18) their characteristic mother-of-pearl lustre are too small to be seen except at high magnifications (Towe and Harper, 1966; Mutvei, 1970); they are of aragonite unless a transformation to the more stable calcite has taken place following fossilization. In prismatic regions crystals may be either of calcite (*Fig.* 19) or of aragonite (*Fig.* 16) depending on the organism. In the crossed-lamellar type, sheets of crystals criss-cross one another as suggested in the outer portion of the shell in *Fig.* 20.

As the preceding figures have indicated, the shells of many invertebrates are a succession of layers, some of the same type in different orientations and others of different types. Many consist of calcite, others of aragonite, and when studying a fossil it is important to determine which is present (Hall and Kennedy, 1967). The outer layers of many Pelecypod shells, for example, are of calcite, whereas the inner ones are of aragonite. Most mollusc shells are of aragonite only, while those of the oysters are purely of calcite.

Aragonite as a mineral is less stable than calcite and tends to invert into it. This has occurred in many initially aragonitic shells and it is important to understand the conditions under which it can take place and to recognize the transformation when it has occurred. At about 500° C. this change is prompt; it is progressively slower at lower temperatures and over geologically significant periods of time it can take place without elevation of temperature. In the dry state it is exceedingly slow, but because aragonite is slightly more soluble than calcite, transformation is greatly accelerated in the presence of moisture. For this reason determination of whether or not aragonite is still present furnishes useful knowledge about the past history of a shell and the probability of its containing undecomposed protein. The two modifications of calcium carbonate can be distinguished by their

*Fig.* 19.—I6954. An obliquely illuminated, polished, and etched section through the calcitic shell of *Ostrea* sp. showing the appearance of prismatic and lamellar regions. (×45.)

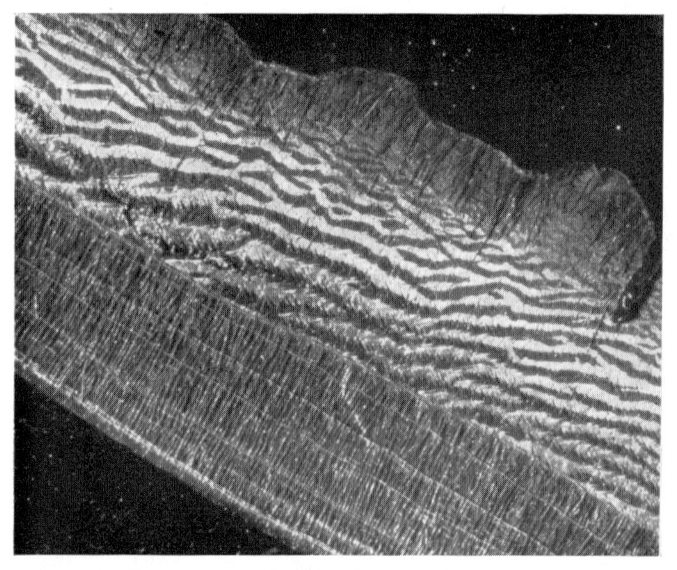

*Fig.* 20.—I7074. An obliquely illuminated, polished, and etched section of the aragonitic shell of *Arca pacifica* showing its differently oriented outer (*upper*) and inner (*lower*) crossed-lamellar structure. (×45.)

reactions to stains and by their somewhat different refractive indices; a simpler and unequivocal distinction can, however, be made on the basis of their unlike X-ray diffraction patterns.

The optical examination of fossil sections serves several purposes. Their patterns of double refraction give the experienced observer a clear idea of the quality of preservation of the mineral structure of bones and teeth. It likewise provides a good estimate of the amount of mineral replacement there has been in all kinds of fossil and supplies a means for the identification of the invading minerals. Recognition of the type of structure prevailing in a fossil shell is a useful guide to further analyses of its composition; according to a recent study (Hare and Abelson, 1965) the amount of organic matter is particularly large in nacreous and prismatic structures, ranging between 1 and 10 per cent. Crossed-lamellar regions in a Recent shell generally are poorer in organic substances, the amount lying between 0·01 and 1 per cent.

As it has done in all other aspects of the study of solids, electron optics has extended in revolutionary fashion the range of what can be learned about the fine structure of fossils. The conventional transmission electron microscope has been portraying details of internal structure down to macromolecular dimensions, and now the same finely focused electron beam is the basis for new instruments to explore both the structure and composition of surfaces. In the electron probe such a beam excites the X-rays characteristic of the material it strikes and thus chemically analyses areas of microscopic dimensions. By sweeping a large area of sample with one of these beams an image is formed, in the scanning electron microscope, which brings out details of either the composition or the contours of its surface.

The individual microcrystals of a bone or tooth were made visible for the first time by the electron microscope. Comparison of their size and distribution in fossil and in fresh specimens thus furnishes an index of alteration far more sensitive than can be achieved through optical microscopic studies. Such comparisons have been made difficult by the need to prepare specimens of an extreme thinness. Adequate sections can, however, be cut by the techniques developed several years ago for dealing with fresh, non-decalcified bones and teeth. In this way, using a diamond knife, the mesh of apatite crystals that constitute the bulk of a tooth or bone can be cleanly sliced without damage. Numerous fossil teeth and bones have now been successfully sectioned, but specimens must be selected with great care because hard minerals like quartz will destroy the edge on the diamond knife.

Examination of these ultra-thin sections has demonstrated that even at high electron microscopic magnifications (*Fig.* 21), the enamel of well-preserved fossil teeth is indistinguishable from fresh enamel. In some of the oldest specimens the microcrystals

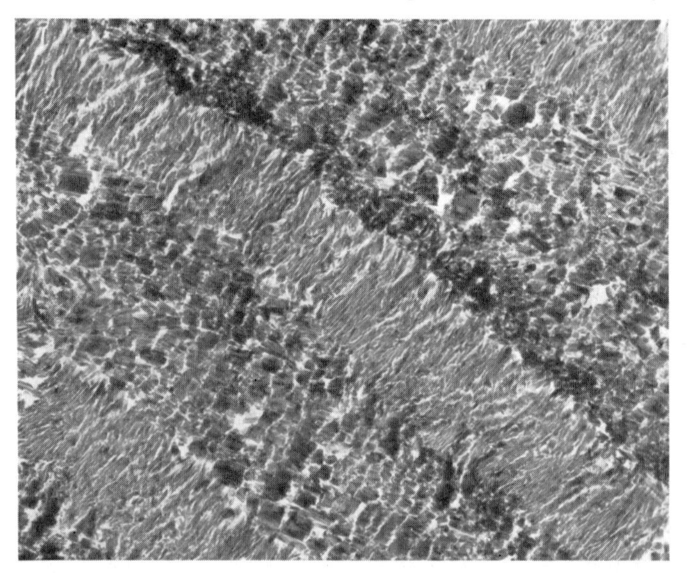

*Fig.* 21.—V633. An electron micrograph of an ultra-thin section through the enamel of a Miocene desmostylus tooth. The enamel rods are the five broad bands of similarly oriented minute crystallites of apatite that run diagonally across the photograph. In adjacent rods the crystals are turned through about 90° to one another. Micrographs indistinguishable from this can be obtained from fresh human and animal teeth. (×9000.)

of apatite are packed in the same way in the individual rods and have dimensions similar to those in present-day teeth. Only rarely has there been evidence that while in the fossil state they have grown, have dissolved and then recrystallized, or that their original arrangement has been disturbed.

Many specimens of fossil dentine and cementum have appeared equally well preserved. Though much of the organic matter they contain when a tooth is alive is lost on fossilization, their apatite may appear unaltered even under the electron microscope. Thus the dentine of a well-preserved fossil tooth will continue to present its original mesh of comparatively well-formed crystallites of apatite. Only in the emptiness of the tubules that radiate through it from the pulp, in their being partially filled with invading minerals, or in the damage to their edges due to the cutting (*Fig.* 22) will it seem different from fresh dentine. Careful study of the region around the tubules frequently reveals even more convincing evidence of the unaltered character of this apatitic

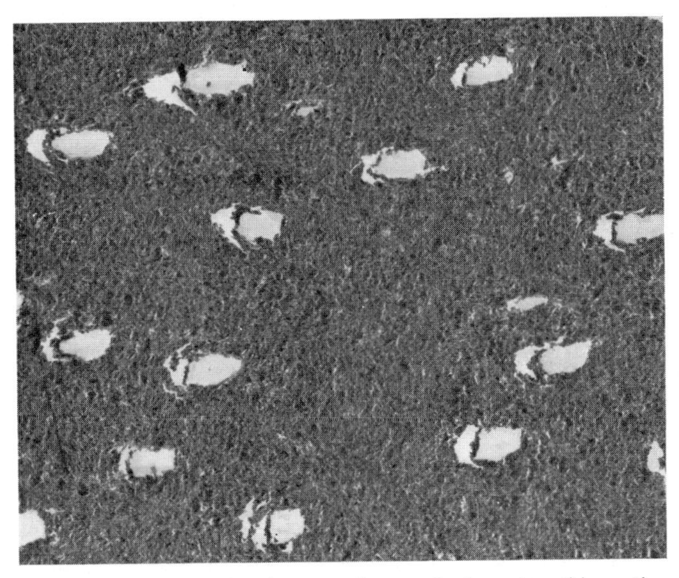

*Fig.* 22.—V63636. An electron micrograph of an ultra-thin section through the dentine of the tooth of a Triassic phytosaur cut normal to the direction of its tubules. Its closely packed but irregularly oriented microcrystals of apatite are clearly visible. In part at least the frayed borders of the tubules are due to damage during cutting. (× 5400.)

meshwork. Under the optical microscope the tubules of a fresh tooth commonly appear surrounded by narrow refractile zones; the electron microscope reveals them as clearly defined regions of microcrystallites still smaller and more densely packed than those making up the body of the dentine. The presence of these peri-tubular zones in fossil dentine, as illustrated in *Fig.* 23, testifies to its unaltered character.

The fact that the microcrystals visible under the electron microscope still have the composition of apatite can be confirmed by electron diffraction observations on the ultra-thin sections. The extreme opacity of matter to electrons and their very short wavelengths make it possible to obtain from minute areas of such a section the diffraction patterns characteristic of a mineral. In this way the crystals in enamel, or the crystalline masses in dentine, repeatedly have been identified as apatite in fossils and fresh teeth. Similar observations on a suspension of a crushed minute fragment of shell can serve to distinguish between calcite and aragonite.

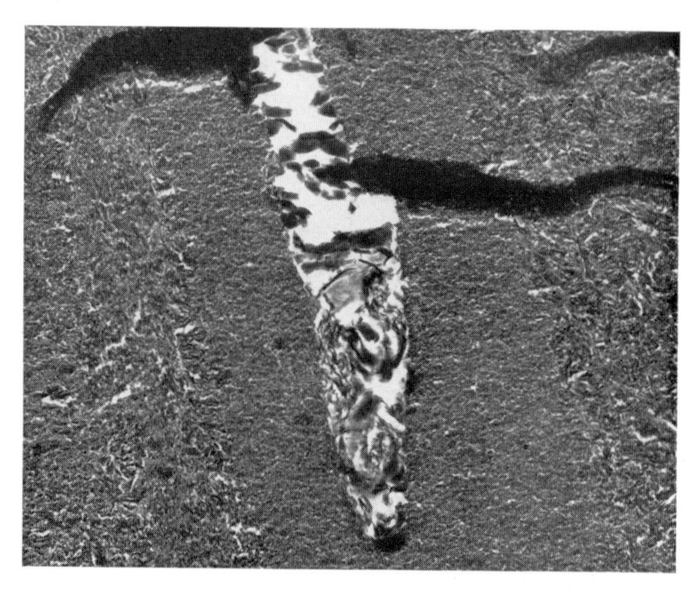

*Fig.* 23.—V6417. An ultra-thin section through the dentine of a tooth of an Oligocene brontotherium cut nearly parallel to the long axis of a tubule. The smaller mean size of the crystallites in the peri-tubular region is apparent. Some of the particles within the tubule are of clay, others may be fossilized bacteria (*see* p. 38). (×7125.)

Living bone, like the dentine of teeth, contains a large quantity of organic matter. In restricted regions its microcrystals of apatite are seen in ultra-thin sections to have the parallel arrangement responsible for the observed optical double refraction but, as with dentine, there is none of the wide-ranging order that characterizes enamel. Individual microcrystals are, however, everywhere visible under the electron microscope and, as in teeth, the quality of fossil preservation can be gauged from their sizes and distribution.

The relatively large size of the calcium carbonate crystals in shells makes it more difficult and less profitable to study them in ultra-thin sections. They, as well as teeth and bones, can, however, be examined by the replica techniques developed years ago for electron microscopy, and often the organic residues exposed by a slow and gentle decalcification can in this way be observed. When knowledge of the surface of a specimen is being sought, it can be replicated directly, but usually it is sub-surface structure that is of interest. This can be revealed by grinding and polishing a surface and then etching it to bring out the sub-surface details. Innumerable procedures have been described for making replicas but for fossils they can usually be made in the following way.

A specimen mounted, ground, and polished as for optical microscopy (p. 6) is lightly etched with dilute mineral or organic acid, the amount of the etch being determined empirically by the character of the specimen and the depth of the detail to be exposed. The etched surface is shadowed by the oblique evaporation on it of an exceedingly thin layer of some high melting-point metal, preferably platinum, followed by the vertical evaporation of a film of carbon. This combined metal–carbon film is covered with collodion and the film thus thickened is freed by dissolution of the specimen in acid. After thorough washing, it is cut into small pieces and picked up on the usual electron-microscope grids. Placed on filter-paper while still moist, acetone is cautiously added in repeated amounts to dissolve away the collodion and leave the replica ready for electron microscopy. It may be a true replica reproducing the contours of the etched surface, or, as a pseudo-replica, it may include bits of the specimen. *Fig.* 24, as an example of this type of replica, shows how accurately it can record details of tubules in the dentine of a fossil tooth. The electron microscopy of such replicas, as *Fig.* 25 illustrates, also permits the most intimate examination of the way the individual crystallites are arranged in a shell (e.g., Watabe and Wilbur, 1961; Heyn, 1962, 1963a, b) and leads to a far more complete understanding than can be gained through optical microscopy of the various types of invertebrate shell structures mentioned on pp. 18, 19.

Organic matter in a fossil is frequently visible as opaque detail in these replicas, but it can be examined more directly as a thoroughly washed suspension of the residue remaining after dissolving a fossil in hydrochloric acid. Many electron microscopic observations of the organic membranes in the shells of several types of Modern invertebrate have shown some that possess little structure, others built of structureless fibres, and a few with

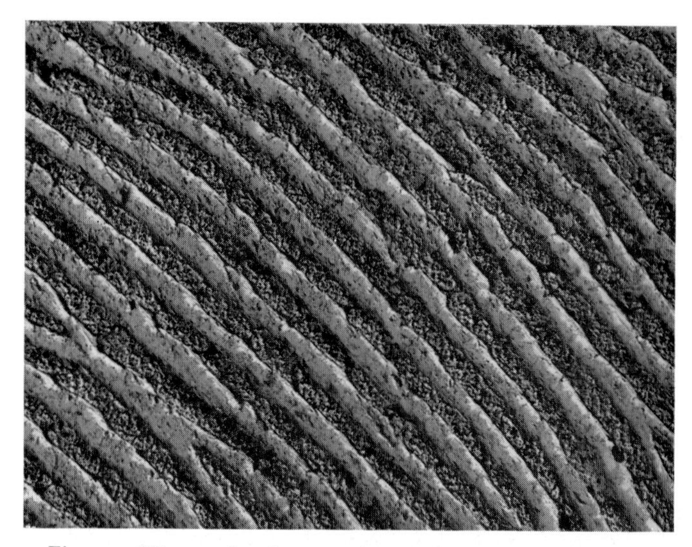

*Fig.* 24.—V6410.—An electron micrograph of a replica, or pseudo-replica, of the polished and etched surface of the dentinal region of a fossil ungulate tooth. Some particles of extraneous mineral present in the tubules have been incorporated into the replica. (×2250.)

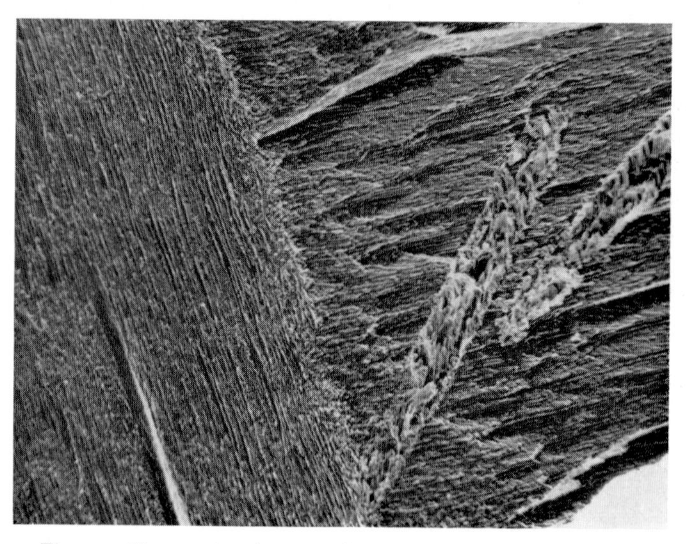

*Fig.* 25.—V7074. An electron micrograph of a platinum-shadowed carbon replica of a polished and etched section through a Recent shell of *Arca pacifica* showing the sharp junction between layers of differently oriented crystals. (×1700.)

fibres having collagen-like striations (Grégoire, 1958a, b; Travis, François, Bonar, and Glimcher, 1967; Travis and Gonsalves, 1969). More than a decade ago (Grégoire, 1959a, b) it was found that membranous fragments from nacreous fresh and fossil *Nautilus* shells viewed in this way have a pebbled fine structure. It is natural to consider the spherical particles responsible for this pebbling as macromolecules of the membranous sheets. At that time the organic, and possibly still protein, nature of some of these fragments was shown by their coloration with the biuret reagent (Grandjean, Grégoire, and Lutts, 1964; Grégoire, 1966a). Such biuret-positive bits of membrane, showing some badly distorted pebbling, were recovered from shells as old as the Ordovician.

Analogous electron microscopic observations made several years ago on the organic residues from fossil bones and teeth (Wyckoff, Hoffman, and Matter, 1963; Shackleford and Wyckoff, 1964; Wyckoff and Doberenz, 1965a, b) revealed microfibrils indistinguishable from those of fresh collagen. Found also in thin sections (*Figs.* 26 and 27) of well-preserved Pleistocene bones and teeth, they had the same distribution as in living tissues and the same sub-microscopic 640 Å striations. Evidently either they were collagen fibres which had remained intact with the passage of time or they were mineral replacements reproducing with remarkable fidelity the original micro details of these fibres. It is the electron microscopic observation of such protein-like structures in fossil shells and bones which has stimulated the chemical work to identify them described in succeeding chapters.

Equally well-preserved striated fibrils have not yet been found in older fossils but unstriated, and even poorly striated fibrils of similar size are frequently seen there (Little, Kelly, and Courts, 1962; Isaacs, Little, Currey, and Tarlo, 1963; Heller, 1965; Armstrong and Tarlo, 1966; Doberenz and Lund, 1966; Pawlicki, Korbel, and Kubiak, 1966).

The newly introduced scanning electron microscope greatly expands the scope of what the conventional electron microscope can do, while avoiding the need to have exceedingly thin specimens. In this instrument a minute beam of electrons scans the face of the sample and either the electrons that are scattered or the excited X-rays are collected by a suitable detector whose output is synchronized with the motion of the electron beam. In this way there is formed on the screen of a cathode-ray tube an

*Fig.* 26.—V63244. An electron micrograph of an ultra-thin section through the femur of a Pleistocene ungulate showing a bundle of the collagen fibres retained within it. The separation between the striae of these fibres is the 640 Å characteristic of this protein. Many of the individual apatite crystals of the bone cap the striae in orderly fashion; others appear as opaque, short, and very thin rods. (×28,000.)

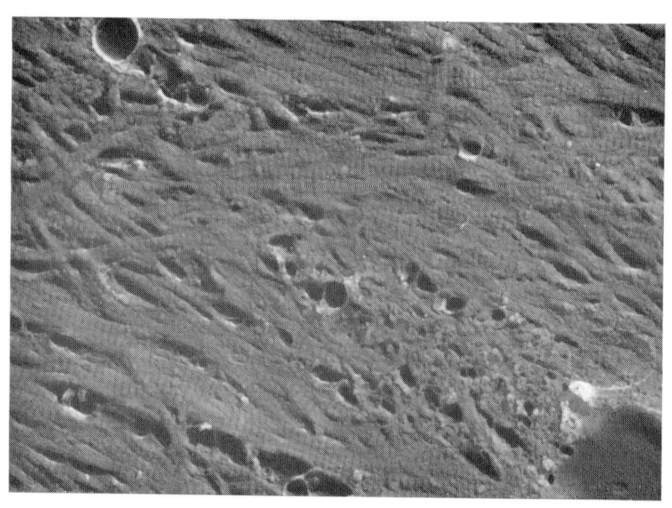

*Fig.* 27.—V64185. An electron micrograph of an ultra-thin section of Pleistocene gopher mandible from which both the apatite and the embedding plastic have been removed and which was then shadowed with platinum. The striated character of the many collagen fibres thus exposed is apparent. (×15,000.)

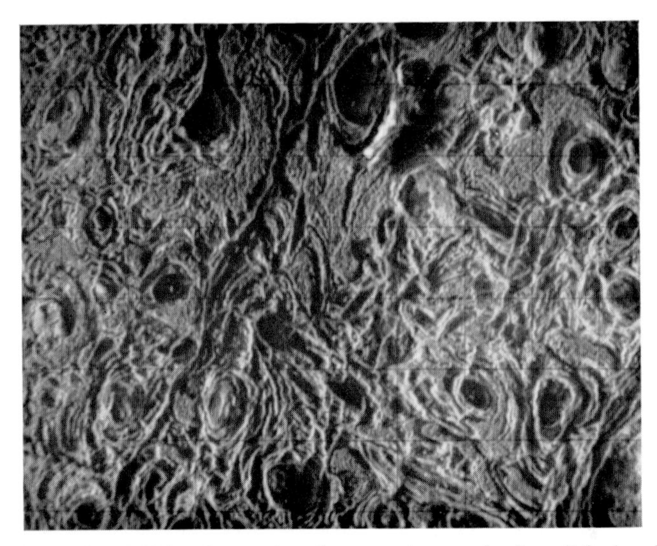

*Fig.* 28.—V6867. A scanning electron micrograph of a polished and etched bone of a Cretaceous member of the crocodile family, for comparison with the optical photograph of *Fig.* 3. (×45.)

*Fig.* 29.—V6884. A scanning electron micrograph of etched enamel of an ancient human tooth, for comparison with the optical micrograph of a similar region of enamel from a fossil desmostylus tooth (*Fig.* 4). (×150.)

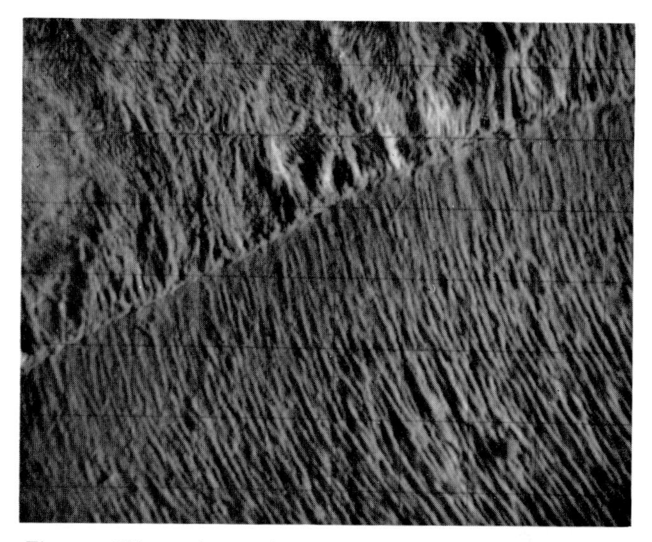

*Fig.* 30.—V6912. A scanning electron micrograph of the dentine–enamel junction of an Oligocene rhinoceros tooth. Dentine with its manifest tubules occupies the lower portion of the field. The smaller enamel rods are seen at the top of the photograph. (×150.)

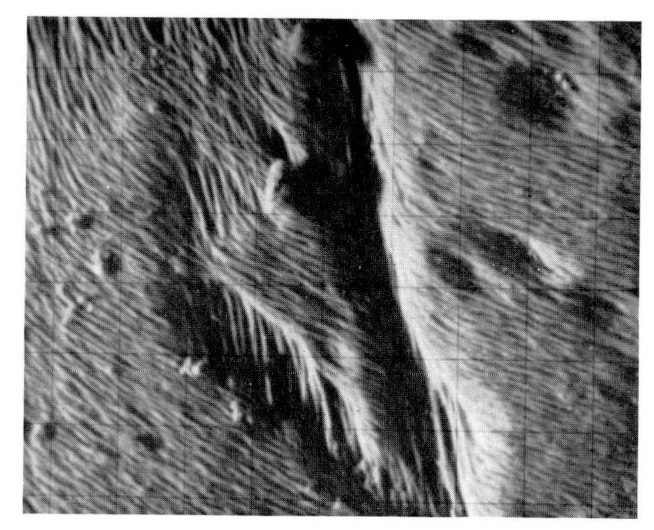

*Fig.* 31.—A scanning electron micrograph of a region of shark fossil dentine. The three-dimensional character of these scanning micrographs is evident. (×90.)

image whose details are determined by the point-to-point intensity of these scattered electrons or X-rays. The X-ray image shows the distribution of a selected chemical element over the face of the specimen; that formed by the scattered electrons has a three-dimensional quality which tells much about the contours of its surface. Attainable magnifications range from those of a simple enlarging lens to the high values reached only by the electron microscope. The outstanding characteristic of the scanning microscope is its ability to form a sharp image of an object that is very far from flat. Though only the surface of a specimen is imaged, internal detail can be revealed by etching and can be photographed at any desired magnification. With fossils this often brings organic material and extraneous minerals into high relief. During observation, electrons are absorbed in the sample as well as scattered; its surface must accordingly have an electrical conductivity sufficient to dissipate the charge that otherwise would develop and give rise to a badly distorted image. When an X-ray image is desired, an adequate conductivity can be obtained by evaporating a very thin carbon film over the etched surface. If scattered electrons are being collected to form an image of surface detail, it is better to coat the surface with a somewhat thicker layer of gold.

Scanning is so recent a development that few fossils have as yet been examined in this way, but several examples will make clear the similarities and differences in the information supplied by optical and scanning electron microscopic images. With this in mind it is instructive to compare the optical picture of bone in *Fig.* 3 with the scanning electron micrograph of *Fig.* 28. At this low magnification they show the same detail and there is little to choose between them. A comparison between *Fig.* 4 and *Fig.* 29 of similar regions of enamel rods suggests the greater amount of information often provided by the sense of depth obtained at higher magnifications from scanning micrographs. *Fig.* 30, of the dentine–enamel junction in a fossil tooth, further illustrates the value of scanning microscopy even at moderate magnifications. *Fig.* 31 is a scanning micrograph of an area of shark tooth for comparison with the dentine of the preceding figure and the electron micrograph of replicated fossil dentine of *Fig.* 24; in considering their relative merits, the greater resolution of the replica is to be balanced against the impression of depth given by the scanning picture.

Scanning electron micrographs will be able to supplement greatly the information to be gained about the fine structure of fossil shells. This is made evident by *Fig.* 32, a scanning enlargement of a small region of the Miocene oyster shell yielding the optical photograph of *Fig.* 19. The richness of structural detail brought out by scanning, even at relatively low magnifications, is

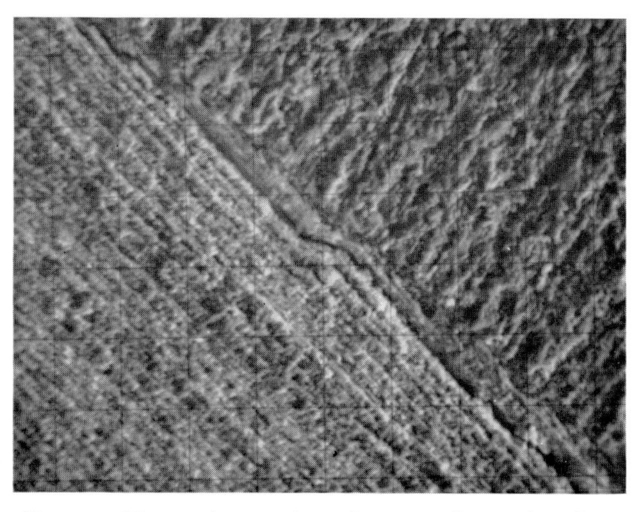

*Fig.* 32.—16954. A scanning electron micrograph taken at moderate magnification of a small region in the Miocene oyster shell photographed in *Fig.* 19. It brings out the unique ability of the scanning microscope to present a wealth of hitherto unseen details of structure. (× 190.)

further illustrated by *Figs.* 33 and 34 made of the Pleistocene *Chione* shell of *Fig.* 16.

The combination of scanning microscopy with probe identification and analysis using the characteristic X-rays generated during examination should be of special value in studying the ancient animals with external hard parts that are partially or wholly phosphatic (*see* Rhodes and Bloxham, 1971). These range from armoured fish and trilobites down to unarticulated brachiopods and conodonts. Through this kind of analysis it is possible to establish the distribution of phosphate throughout a fossil without destroying it and, where phosphate is high, to identify the compound of which it is part by measuring the Ca/P ratios.

Most of palaeontology has been built around the study of fossilized hard tissues, but mummified and fossilized specimens of

*Fig.* 33.—I7080. A scanning electron micrograph of a fragment of the Pleistocene *Chione gnidia* shell of *Fig.* 16. The white patches are areas damaged by dirt and abrasion. (×70.)

*Fig.* 34.—I7080. A scanning electron micrograph taken at a somewhat higher magnification of an area of the shell of *Chione* shown in *Fig.* 33. Much fine structure is beginning to be visible. (×190.)

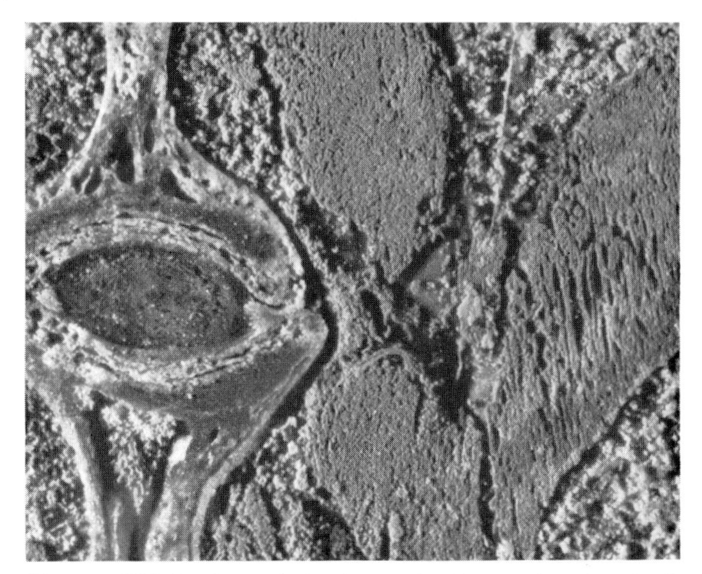

*Fig.* 35.—V6828. An optical micrograph of a portion of a fossil Jurassic fish showing a vertebra and associated 'muscle'. (×45.)

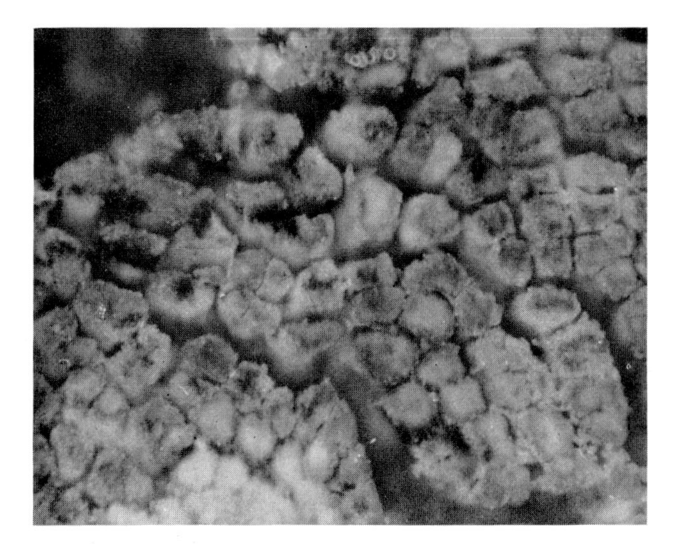

*Fig.* 36.—V6828. An optical micrograph of a thin section through one of the masses of fossil 'muscle' of *Fig.* 35, showing the non-uniform character of the sectioned filaments. (× 190.)

soft tissues are occasionally found. Sometimes cellular structure is visible in the contents of a fossilized marine or insect shell, in the open spaces of a bone or the cementum of a tooth (*Fig.* 11). Most fossil skin is certainly no more than an imprint of the original, but specimens have been uncovered in which fossil skin appears to be enclosing something more than bones. The techniques now at our disposal may be expected to reveal much that

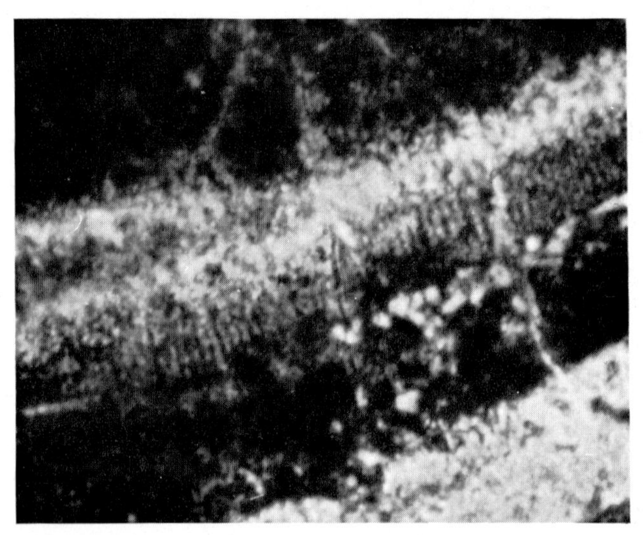

*Fig.* 37.—V6828. Another field of sectioned 'muscle' of the fossil fish of the preceding two figures. The fibre across the centre is striated with a period corresponding to that found for modern striated muscle. (× 1065.)

is new about the original structure and the organic matter persisting within such objects.

An idea of the minimum that can be learned through these studies is provided by examination of the remains of a fossil fish of Jurassic age (Wyckoff, 1971). As *Fig.* 35 indicates, masses of what grossly appear to be muscle are attached to its backbone. Internal detail in the individual fibres of this material is apparent at the higher magnification of *Fig.* 36; proof that it is indeed mineralized muscle lies in the fact that some of the fibres of a bundle still show, under the optical microscope (*Fig.* 37), the periodic banding typical of present-day striated muscle.

In this Jurassic fish the fossil muscle was readily exposed for study. Often, however, we may wish to find out if interesting

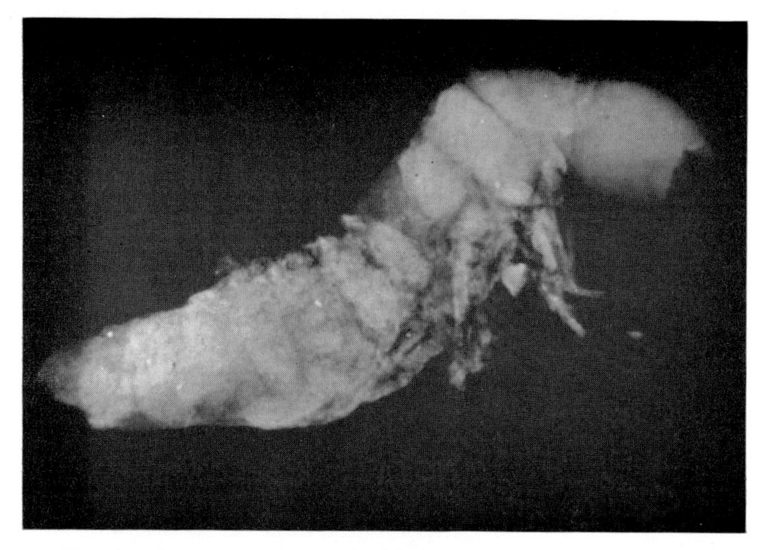

*Fig.* 38.—A microradiograph of a Miocene fossil insect, *Schistomerus californense*, showing much internal detail. (×90.)

*Fig.* 39.—An optical micrograph of the Miocene insect of *Fig.* 38 showing the excellence of preservation of its shape and surface details of structure. (×90.)

detail is contained within a specimen too thick or opaque for optical examination. For many of these, X-ray radiography, as employed in dentistry and medicine, can aid (e.g., Stürmer, 1966). The amount of information to be gained will, of course, depend on the difference in opacity to X-rays of the specimen and its enclosing rock. It often is not great for a well-preserved bone or shell, but if there has been massive replacement by an iron-rich mineral, as with the foregoing muscle, much can frequently be seen in a radiograph of a still-embedded fossil. When fossilization has taken place in a swampy environment it is not uncommon to find soft parts of an animal replaced with great faithfulness by pyrite ($FeS_2$). Pyritized fossils are common in coal deposits (for instance Ehlers, Stiles, and Birle, 1965) and in organic-rich shales and slates. A very instructive example of what can be seen within these pyritized fossils is provided by radiographs of cephalopods and trilobites in Devonian slates (Stürmer, 1970).

The structures to be seen by simple radiography are necessarily coarse, but by employing the techniques of microradiography it is possible to reveal detail of microscopic dimensions. This involves employing an X-ray source having an unusually small focal spot. Radiographs made with it are essentially pin-hole photographs; their resolution is comparable with the size of the spot. X-ray tubes that are commercially available or can rather easily be made in the laboratory have focal spots a fraction of a millimetre in diameter. Even smaller X-ray sources have been made by building a tube in which the electron beam is reduced in cross-section by electron lenses (as in an electron microscope) and focused on a thin metal foil which acts as target (e.g., Cosslett and Nixon, 1960). *Fig.* 38 is a microradiograph (Doberenz, Matter, and Wyckoff, 1966) of a minute fossil insect of the same species as that shown in the optical micrograph of *Fig.* 39. Many details of internal, and otherwise invisible, structure are to be seen in these X-ray pictures of microfossils. Their surface detail can be portrayed by scanning and their composition determined by probe analysis. In this fashion it was shown that part of the insect fossil of *Fig.* 38 consisted of celestite ($SrSO_4$) and the rest of quartz ($SiO_2$).

Demonstration that soft tissue does indeed become fossilized with the preservation of its external form and at least some of its inner structure is especially important because the primordial

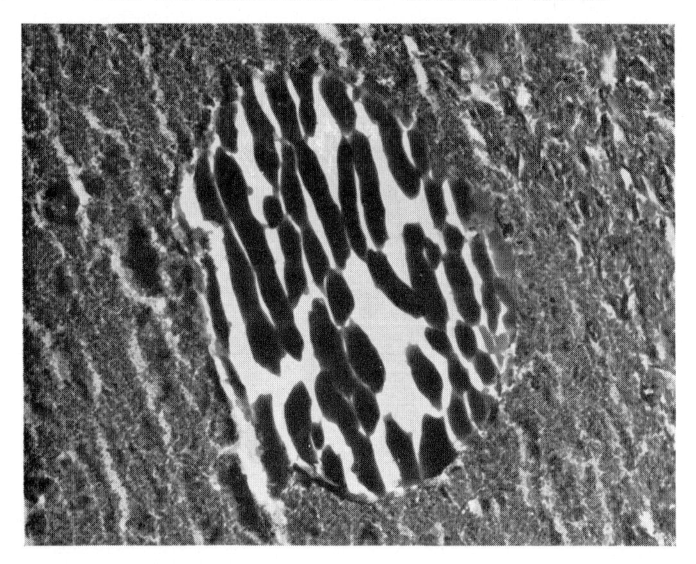

*Fig.* 40.—V6417. An electron micrograph of a section transverse to that of *Fig.* 23 showing a dentinal tubule of an Oligocene brontotherium tooth. The objects filling the tubule have the dimensions and general appearance of bacteria. ($\times$ 15,000.)

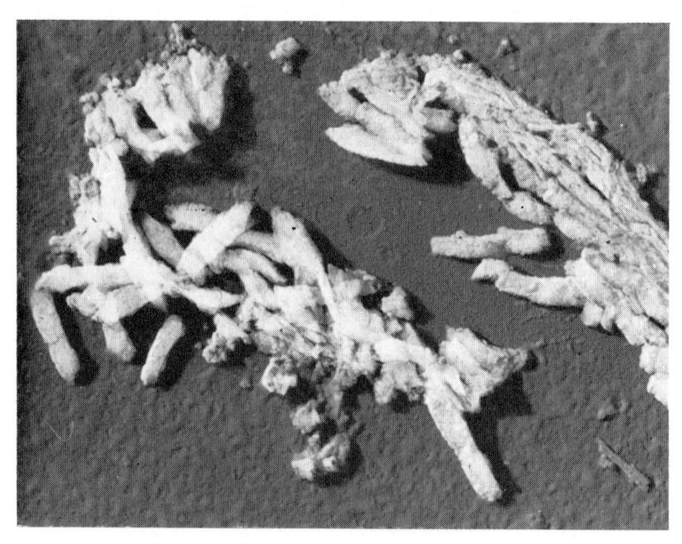

*Fig.* 41.—V6417. An electron micrograph of a section similar to that of *Fig.* 40 after removal of the plastic, dissolution of the apatite with acid, and shadowing with platinum. Several of the bacterium-like bodies occur in pairs that suggest division. ($\times$ 20,600.)

forms of life must have been masses of soft tissue unprotected by the hard coverings later developed by insects and other invertebrates. Except for algae, some as old as the Precambrian, few fossils are known which can unreservedly be accepted as belonging to this earliest palaeontological record. Objects considered to be fossil bacteria were first described almost a hundred years ago, and at the turn of the century they were extensively studied as seen in fossil bones, teeth, and deposits associated with coal and sedimentary rocks (Renault, 1899). Renewed attention was given them at about the time of the First World War (Walcott, 1915a, b; Moodie, 1920). Much of the evidence then presented for identifying them as micro-organisms was necessarily based on the pathological changes they were supposed to have caused. The resolution of the optical microscope, however, was not sufficient to establish convincing connexions between bone and tooth damage and the presumed bacterial agents, and interest in the subject flagged. Now the great resolving power of our new techniques calls for a fresh examination of what, if real, will include the fossil remains of the most primitive forms of life. Several papers dealing with them have recently appeared (for instance, Marshall, May, and Perret, 1964; Meinschein, Barghoorn, and Schopf, 1964; Barghoorn and Schopf, 1965, 1966; Schidlowski, 1965; Schopf, Barghoorn, Maser, and Gordon, 1965; Jackson, 1967; Dennis, 1969). The evidence is not yet conclusive and there is now the opportunity to re-examine their association with demonstrably diseased bones and teeth.

The scanning electron microscope with its ability to portray fractured but otherwise undisturbed surfaces at high magnifications is particularly fitted to contribute to this study and much can be expected from its use. A continuing scepticism about the preservation of even the shape of organisms which are little more than soft lumps of protoplasm is healthy. The kind of evidence we must rely on to overcome it is illustrated by *Fig.* 40, which shows some fossilized 'bacteria' in the dentinal tubules of an Oligocene brontotherium (Doberenz and Wyckoff, 1967a). The likeness to a cluster of bacteria becomes more striking (*Fig.* 41) in an adjacent section, metal-shadowed after removal of apatite and plastic.

It is scarcely conceivable that soft tissue will have preserved its original composition, but it is essential to ascertain if any organic matter remains and what minerals have replaced the protoplasm. When the fossil object is observed in thin section, a transparent replacing mineral, such as quartz or calcite, can usually be identified by the customary petrographic procedures. If the mineral is opaque or if the detail in question is superficial or can be exposed by etching, the electron probe is a powerful tool for its identification.

In this probe an electron beam of minute cross-section, like that employed in the scanning microscope, impinges on a chosen area of the sample and the excited X-rays are analysed with an X-ray spectrometer. From estimates of the relative intensities of the characteristic radiations of the chemical elements in the irradiated spot, its mineral composition can ordinarily be ascertained. Such a probe examination of the fossil muscle of *Figs.* 35 and 36, for example, would show the dominance of the K lines of iron and silicon, corresponding to the fact that the original muscle tissue had in this instance been replaced with great faithfulness by an admixture of quartz and limonite. If the spectrometer was one that could deal with very long X-rays, it would register the oxide character of the replacement and would show if carbon from the original tissue had been preserved. The persistence of carbon could be determined in other ways as well. For instance, a bit of the fossil muscle could be examined under the optical or scanning microscope before and after heating in air, or by heating the specimen in oxygen or hydrogen and ascertaining with a gas chromatograph if carbon dioxide or hydrocarbons (p. 48) were evolved.

In the preceding discussion emphasis has been placed on the investigation of internal structure as an index of the quality of preservation of a fossil. The deeper penetration made possible by electron optical instruments has, however, done more than improve our ability to select the best specimens; by showing that structure is often preserved down to macromolecular dimensions, it has opened up new perspectives to those seeking to trace the course of evolution. Not only has evolution involved an increasing complexity of organization, but it seems probable that the living matter thus organized may itself have become more complex. With the help of the sensitive and powerful methods of examination and analysis now at our disposal, a search can begin for evidence bearing on this question. The following chapters are devoted to some of the results obtained while thus studying the chemical composition of fossils, including the organic remnants they have retained.

## *CHAPTER III*

# THE COMPOSITION OF FOSSILS

THE demonstration that the internal structure of many fossils has remained substantially unaltered with the passage of time makes it worth while to learn everything possible about their present composition. There are two aspects to such a study and different techniques are required to deal with each. One aspect is concerned with the inorganic framework of the fossil, the other with whatever organic substances may remain as residues of its living matter.

The investigations of fine structure discussed in the preceding chapter make it certain that many well-preserved fossil bones, teeth, and shells retain intact their original phosphate and carbonate frameworks. Though predominantly calcium compounds, these hard tissues commonly possess trace amounts of other metallic atoms and, in the case of shells, they may contain more than a trace of magnesium. A complete study of a fossil will accordingly include an analysis for trace elements and, when found in unusual amounts, an attempt to explain their presence. They may be derived from minerals which have entered the fossils to take the place of decayed organic matter; as such they give useful information about what has happened during the long period of fossilization. On the other hand, they may be native to the original hard tissue; they then can be a source of knowledge about conditions under which the animal lived and died.

A variety of techniques now exist for the rapid, routine determination of trace elements. One is activation analysis, which consists in exposing the specimen to an intense neutron bombardment followed by a radiochemical analysis for the chemical elements created in this way. Some elements, such as arsenic, can thus be recognized in amounts of less than one part in a billion; no such sensitivity is achieved for many others. Despite its advantages in particular instances, complete activation analysis is so costly and time-consuming that it will probably only rarely be applied to fossils.

X-ray spectroscopy is a simpler and more practical means of establishing minor constituents and trace elements. Though other procedures detect smaller amounts of certain elements, a routine X-ray spectroscopic scan gives semiquantitative information about all but the lightest chemical elements, automatically and in a few minutes. The minimum amount of an element it can directly reveal ranges from about 70 to about 10 parts per million, but if necessary this sensitivity can often be enhanced through preliminary chemical fractionations. When needed, recourse to especially prepared standards will yield more accurate results. It is one of the great advantages of this form of spectroscopy that in its routine use no special specimen preparation is needed. A powdered sample is exposed to X-rays and the resulting fluorescent radiations characteristic of the chemical elements present are identified on a spectrometric chart. Thus the procedure is non-destructive and the sample can subsequently serve for other analytical determinations. Unlike optical spectra, the X-ray spectra are similar for all the elements. Each consists of but two (or three) lines strong enough to require consideration; the wavelenths of these few lines, increasing in a regular fashion with decrease in atomic number, lead to direct identification of the element.

Results are so immediately forthcoming that in this laboratory a preliminary X-ray spectroscopic scan is made of every fossil considered for study. This serves a dual purpose: it reveals trace elements that may be of interest and at the same time gives a valuable indication of the amount of secondary mineral replacement that has occurred. With certain fossils it has proved helpful to supplement the information gained with X-rays by optical spectroscopic examination. For some, but not all, metals this is the more sensitive technique; however, the hundreds and even thousands of lines in the spectra of many metals make reading them time-consuming, and for survey purposes there is the further disadvantage that electronegative elements fail to give useful spectral emission in the optical region.

Recently there has been developed an atomic absorption spectroscopy which is quantitative and has great sensitivity for many metallic elements. It cannot be employed for surveys to ascertain what trace elements may be present, but it is of great value in the analysis for certain elements that are minor constituents in fossils. In this type of spectroscopy a solution of the sample is aspirated into a flame hot enough to create an atomic vaporization. Light from a special lamp emitting the spectrum of the element in question is passed through the flame and a sensitive line of this spectrum is isolated by passage through a spectrometer. The neutral atoms produced in the flame absorb energy from the light and the resulting reduction in intensity of the line selected for measurement is a function of the concentration of the element in the solution.

The occurrence of magnesium as a minor constituent of shells has been studied extensively in the past using conventional chemical methods (Clarke and Wheeler, 1922). It can be determined by X-ray spectroscopy but with far greater ease and sensitivity by atomic absorption; as little as 1 part in a million can thus be quickly established. There is a direct relation between the quantity of magnesium in a fossil shell and its original crystalline form; the determination of this element is therefore made on every specimen with which we deal. Analysis is carried out by introducing a dilute acid solution of the fossil into the flame of the absorption spectrometer equipped with a magnesium hollow-cathode tube. By replacing the magnesium tube with one emitting calcium radiation, the calcium content of the shell can be determined in the same solution and this provides, by difference, an immediate determination of the total quantity of foreign minerals in the fossil.

The chemical analyses for magnesium made on the shells of hundreds of different living invertebrates have shown that much more of this element is found in calcitic shells than in those whose calcium carbonate is in the form of aragonite. The lesser amount of magnesium in aragonitic shells is readily understood from the fact that magnesium carbonate, possessing the atomic arrangement of calcite, does not form extended solid solutions with aragonite. In contrast, magnesium atoms easily replace those of calcium in the isomorphous calcite structure. The early chemical analyses indicated that magnesium-containing calcitic shells fall into two different groups: those that contain less than about 2 per cent, pelecypods and gastropods, and those containing 8 or more per cent. Among the latter are calcareous algae and certain echinoderms. Very few specimens lie in the zone between. It has been natural to wonder if high-magnesium shells contain dolomite, $CaMg(CO_3)_2$, but a few X-ray powder photographs made a generation ago gave only calcite-like patterns. The differences between the calcite and dolomite patterns, though definite, are minor, and in view of recent improvements in technique, a further X-ray study of magnesium-rich shells seems desirable.

A determination of the magnesium content of a fossil shell is of value both as an indication of the conditions under which the animal lived (Stanton and Dodd, 1970) and for the evidence it gives as to whether or not a shell now calcitic may originally have been of aragonite. As already pointed out (p. 19), aragonite frequently transforms with time into the more stable calcite. It is therefore important to ascertain, for instance through careful microscopical examination, if calcitic, magnesium-poor shells have undergone such a transformation. The amount of magnesium that enters the calcite arrangement must depend on both its concentration in the water in which the shell has grown and on the temperature. In well-preserved fossils where no recrystallization has taken place, the quantity should therefore be an indication of this temperature. Though little use seems thus far to have been made of this indicator of earlier marine temperatures, future work may be expected to test its value, especially employing the many shells in which both calcite and aragonite are present as distinct layers. Magnesium occurs in vertebrate as well as invertebrate fossils, but there has been no extensive study of its distribution in the former.

Strontium (Noll, 1934) is another element present as a minor constituent of fossils, both invertebrate and vertebrate. Though early analyses were by chemical methods, it can best and very simply be determined by X-ray spectroscopy. Atomic absorption can equally well serve to establish the strontium content in shells, but care must be taken when applying this technique to bones and teeth. Very high flame temperatures are required to dissociate the phosphates and reliable results with them are obtained only through the addition of some cation such as lanthanum, which liberates calcium and strontium by forming with phosphate an even more thermally stable substance.

Strontium is present in nearly every fossil bone and tooth but in widely differing amounts ranging up to about 2 per cent. The quantity of this element in soil and rocks varies considerably with the geographic region and there is a correlation between it and the amounts present in the teeth and bones of animals living in a region (Thurber, Kulp, Hodges, Gast, and Wampler, 1958; Wyckoff and Doberenz, 1968). Undoubtedly this is to be attributed to its uptake by way of the plants that serve as primary sources of food (Toots and Voorhies, 1965).

This element, like magnesium, is a minor or trace constituent of most invertebrate shells (for instance, Lowenstam, 1964a, b; Hallam and Price, 1966; Price and Hallam, 1967). It is found in relatively large amounts (up to about 0·5 per cent) in aragonitic shells, in smaller quantities (to 0·1 per cent) in those of calcite. In this respect its occurrence is the reverse of that of magnesium. This is due to the fact that strontium carbonate and aragonite, with the same crystal structure, can form a broad range of solid solutions. Calcite, on the contrary, cannot take many large strontium ions into its different atomic arrangement.

As is true for magnesium, the amount of strontium incorporated into a shell depends on the amount in the sea in which the animal is living and on the water temperature. *Mytilus* shells are among those that contain both aragonite and calcite as separate layers. It has been shown (Dodd, 1965) that in them the content of strontium in their aragonite layers falls with rise in temperature but that the small quantity in the calcitic layers increases. This produces demonstrable seasonal fluctuations in a single shell. Recently these relationships have been used to estimate the temperature under which a group of Pliocene and Pleistocene *Mytilus* shells were formed (Stanton and Dodd, 1970). In making such an estimate it is assumed that the Ca/Sr response of these ancient invertebrates is the same as that of their Modern relatives and that the composition of the fossil shells has remained unaltered. The agreement between the temperatures thus obtained and those given by other methods of estimating palaeotemperatures indicates that Ca/Sr ratios are indeed useful indices of ancient temperature relationships.

A number of other chemical elements have been noted in considerable amounts in many fossil bones and teeth and it is highly probable that additional elements in trace amounts would be found if a systematic search were carried out. Those that have been encountered during the routine X-ray spectroscopic surveys made in this laboratory have been related to specimens from a particular geographic region, and even from a particular fossil site. Probably most have entered the fossil as a secondary replacement mineral; as such they have little to say about the initial state of the bone or tooth. When an element is found in both the fossil and its matrix, this is evidently the case. There have, however, been instances where it is not true; then it must be assumed either that the element was present in the living bone or that it was taken up from the surroundings after death but before it was enclosed in the rock in which it is found.

A somewhat surprising feature of these observations has been the frequent occurrence of the uncommon element yttrium (Matter, Davidson, and Wyckoff, 1970; Parker and Toots, 1970), especially since it has rarely, if ever, been accompanied by chemically related elements such as the rare earths. It is present in measurable amounts, up to more than 0·2 per cent, in perhaps a third of the specimens from certain localities in both the Old World and the New. Rarely has it been detected in a matrix. The significance of this finding is not yet clear, but it is to be noted that yttrium forms a highly insoluble phosphate and as a radio-isotope can accumulate in bone (Jowsey, Sissons, and Vaughan, 1956; Jowsey and Orvis, 1967). Yttrium phosphate could conceivably become concentrated in a bone, but it remains to be determined if this has happened during life or after fossilization.

In several instances one or more unusual elements have been found in a large percentage of the fossils from a given locality (e.g., Matter and others, 1970; Wyckoff, 1971). Lead and arsenic have thus been observed in bones but not in the surrounding rocks. It is intriguing, but at present rather futile, to speculate as to whether or not the accumulation of animal remains in these localities is related to the toxicity of these elements. In other deposits uranium or thorium has been a common trace constituent of the recovered fossils. It is well known that these elements become concentrated in organic matter exposed to their weak solutions, and probably this is the way they entered the fossils. When, as has sometimes been true, they have been completely absent from the enveloping rock, this absorption must have antedated enclosure in the matrix in which they have been found. Observations such as the foregoing are yet too fragmentary to have much bearing on important palaeontological problems, but as they increase in number they are bound to show a mounting relevance.

There are many physical techniques in addition to those described here which can tell about the past history of a fossil. One of the most valuable is the determination of the isotope ratios for certain of its chemical elements. Where an isotope of an element is radioactive, measurement of this activity and of the product of its decay is a useful way to determine the age of a sample. Lead isotope, rubidium–strontium, and potassium–argon ratios are constantly employed to establish the age of ancient rocks; $^{14}C$-dating is applicable to fossils that are late Pleistocene and younger in age. Ratios of stable isotopes, of $^{18}O/^{16}O$ and of $^{13}C/^{12}C$, depend on environmental conditions, especially temperature, prevailing when a fossil bone or shell was formed; their determination consequently makes an important contribution to its history. All these are well-established techniques whose applications have frequently been discussed in books on geochemistry. The results they provide have only a peripheral bearing on the central objective of this monograph which is to examine and identify the organic residues remaining in individual fossils.

Investigation of the inorganic part of a fossil lays a necessary foundation for the more significant, careful study of these organic residues. Such a study must be carried through in a series of steps, the first of which is often an estimate of the content of organic carbon. This serves two purposes. It weeds out specimens too poor to merit further consideration and at the same time indicates the size of sample required for analysis of the various types of compound present. It can be done, as already mentioned in Chapter II, by burning a sample in oxygen and measuring the amount of carbon dioxide formed, or by a pyrolysis in hydrogen followed by determination of the resulting hydrocarbons.

Shells consist of calcium carbonate and bones often have calcite as a replacement mineral. If, therefore, the organic matter of a fossil is to be burned, it must be done under conditions which do not mix the carbon dioxide thus produced with products of carbonate dissociation. When volatile hydrocarbons can be ignored, the quantity of other organic carbon can be found by first heating the sample in helium to a temperature high enough to decompose carbonates. Combustion can then be effected by transferring to oxygen and collecting the products of combustion which are passed successively through a drying column (to remove the water that has been formed) and a U-tube cooled with liquid oxygen to condense the carbon dioxide. After clearing the system with helium, the carbon dioxide is evaporated and its amount determined by passage through a gas chromatograph. When soluble amino-acids and peptides can be ignored, it will be found simpler to decompose carbonates with acid before heating the insoluble residue in oxygen.

Pyrolysis provides a rough estimate of the total non-volatile organic carbon. It is carried out by heating the sample for a few seconds in a mixture of helium and hydrogen and measuring the resulting hydrocarbons in a gas chromatograph. Several hydrocarbons yielding characteristic groupings of chromatographic peaks often result from the pyrolysis of an organic compound. Experiment has shown, however, that a fossil can be pyrolysed at a temperature at which carbonate is not dissociated, graphite does not react, and practically all the hydrocarbons formed are concentrated in a single peak whose height is an adequate indication of the organic content.

The content thus estimated may be a wide range of compounds derived from living matter. Some may be the remains of animal life, others of plant life; still others, like the fatty acids, proteins, and carbohydrates, may be common to both. Fossil plant material has been extensively studied as the source of coal and the presumptive source of petroleum. For the most part it consists of cellulose, lignin, chlorophyll, and products of their degradation. Some, like the porphyrin derivatives, are identifiable compounds; others, like kerogen and humic deposits, are chemically ill-defined materials which under drastic chemical treatments can be broken down into a variety of simpler substances. They have been studied *en masse* and not in relation to individual fossils, and their consideration falls outside the scope of this monograph.

Fossil animal remains, apart from the inorganic carbonate and phosphatic hard tissues already discussed, are almost exclusively derived from the proteins and lipids that are essential components of all living matter. The fats as such are rarely preserved in their native state, but they yield the stable fatty acids found in even very ancient fossils. Modern proteins are highly individual complexes built up of the same twenty-odd amino-acids in various proportions and sequences. They reflect the innumerable differences that exist between living forms. If evolutionary development has been accompanied by increases in the complexity of living matter, evidence of it is most likely to be found in the proteins. It is this fact that gives special importance to the presence in fossils of proteins and of the peptides and amino-acids that result from their degradation.

So little of the original organic matter is preserved in most fossils that only recently developed methods of chemical analysis have the sensitivity needed to do more than detect its existence. It is chromatography in its various forms which has put us in a position to make detailed quantitative analyses of the minute quantities of proteinaceous and lipoidal substances recoverable from fossils of great age. In the first applications of chromatography to proteins, the amino-acids obtained by hydrolysis were separated one from another by taking advantage of the different rates at which they could be made to migrate over filter-paper. This paper chromatography and its subsequent refinements permitted identification of the amino-acids in a protein and a rough estimate of their relative amounts. Its value, however, ceased when later techniques based on liquid–column separation began to yield quantitative data. These techniques make use of cation-binding resins such as Dowex 50 or Amberlite 120 which are able to take up all the amino-acids through their amino ($NH_2$) groups. When buffers of appropriate *p*H slowly move through a resin column upon which a mixture of amino-acids has been placed, the acids will be released one after another to appear in separated packets in the liquid issuing from the column. By reacting this effluent continuously with the ninhydrin reagent that produces a strongly coloured product with each acid, successive maxima of light absorption give rise to chart peaks whose areas accurately measure the amount of each amino-acid. This is the basis for the automatic amino-acid analysers which are now in use in numberless hospitals and laboratories throughout the world.

In early instruments, relatively large samples were needed for a satisfactory analysis and the flow was so slow that nearly 24 hours were required for all the amino-acids to emerge from the column. By improving the resins and employing high pressures, sensitivities have been increased and the time for an analysis has been reduced to no more than 2 hours. Results obtained by this liquid–column chromatography of fossils are discussed in the next chapter.

Gas chromatography makes use of the fact that when a mixture of gases is introduced into a stream of helium or nitrogen flowing through a capillary column filled with impregnated diatomaceous earth, the components of the gas mixture will be retarded for different lengths of time. Appearing one after another in the effluent, the amount of each can be measured by suitable detectors. By maintaining the column at an elevated temperature, solids and liquids that vaporize without decomposing, as well as gases, can be analysed in this fashion.

The fatty acids found in larger or smaller amounts in most fossils are readily determined by gas chromatography. Many are volatile enough for direct analysis, but to ensure that all will be recorded, they are usually converted into their more volatile methyl esters before analysis.

The amino-acids do not vaporize without decomposition, and it has been harder to find for them appropriate volatile derivatives upon which a practical method of analysis can be based. The reactions producing such derivatives must be simple, rapid, and complete for every amino-acid in a mixture of all. Among those that have been tried, the reaction worked out by Gehrke and his associates (Gehrke, Roach, Zumwalt, Stalling, and Wall, 1968) has been found suitable for analysing the small quantities present in fossils. It involves the simultaneous conversion of a mixture of amino-acids into their *N*-trifluoroacetyl-*n*-butyl esters. These compounds differ considerably from one another in volatility and they emerge one after another from a chromatographic column whose temperature is being raised in a carefully controlled fashion. They are conveniently measured in an instrument equipped with a hydrogen flame detector; in it, hydrogen and air mixed with helium carry the vapours to be analysed through the apparatus and this mixture is burned in a controlled way as it emerges from the column. The detector measures the electrical conductivity of the flame which sharply alters whenever the burning mixture contains the vapour of one of the components of the injected sample; in consequence the output of the detector appears as a series of peaks whose areas are proportional to the quantity of each component. As in liquid–column chromatography, the peaks are identified and their areas evaluated by reference to runs with standard mixtures of the pure components.

Comparison with liquid–column chromatography has demonstrated that this type of gas chromatography can be successfully applied to fossil hydrolysates and that the quantity required for analysis is far less (Akiyama, Davidson, Matter, and Wyckoff, 1971). This advantage is partially offset by the fact that substances present in the fossil hydrolysates may interfere with the complete conversion of the amino-acids to their esters and may themselves react to yield extraneous peaks difficult to identify. As results to be described in the next chapter will suggest, the two kinds of chromatography, however, work together in greatly extending the range of fossils that can be successfully analysed.

## *CHAPTER IV*

# THE PROTEINACEOUS RESIDUES OF FOSSILS

PROTEINS as a class are chemically unstable. Minor molecular changes occur when they lose their most characteristic properties, such as their ability as enzymes to catalyse specific reactions or as respiratory proteins to transport oxygen. Usually such denatured proteins have lost whatever solubility they may originally have possessed. As alteration proceeds, their very large, three-dimensional molecules commonly break up into string-like polypeptide fragments and these in turn split to yield amino-acids (whose linked residues make up a polypeptide). Unlike the proteins they form, amino-acids are stable chemical compounds and it therefore is not surprising that they are found as the final residues of the proteins which the fossils originally contained. It has been known for some years that they can be recovered both from fossil shells and from some ancient sediments.

These free amino-acids are convincing evidence that proteins originally were present, but their identification and quantitative determination give only very limited knowledge of the composition of the protein from which they were derived. This is due in part to the fact that the amino-acids are soluble in water and in the kinds of saline solution that may at one time or another have percolated through a rock and its enclosed fossil remains.

What we find are only those compounds that happen not to have been leached away during the course of hundreds of thousands or millions of years. Though most of the amino-acids that enter into the composition of proteins have considerable chemical stability, this stability is not unlimited under the geological conditions to which a rock and its enclosed fossils may have been subjected. Furthermore, they differ amongst themselves in their response to heat and probably to pressure. Some will accordingly be much longer-lived than others and this, too, is a factor influencing the free amino-acid content of fossil specimens.

Most proteins are so unstable that it has generally been taken for granted that they could not persist for geologically significant periods of time, and until recently little serious effort was made to recover them from fossils. When, however, the electron microscope revealed protein-like macromolecular particles in certain fossil shells and seemingly intact collagen fibres in Pleistocene bones, it became urgent to establish if these objects were still the proteins they appeared to be or were artefacts composed of replacing minerals. Then, once it had been found that they were indeed organic in composition, the detailed analyses of these residues became a possible source of new information about the course of evolution.

Before proceeding to review the results obtained through such analyses it may be worth while to describe somewhat more fully the chemical nature of proteins and their decomposition products, and the way their analyses are carried out. The unit of protein structure, an amino-acid, is an organic compound which has an amino, $-NH_2$, and a carboxyl, $-COOH$, group as part of a single molecule. The carboxyl group is rendered acidic through the dissociative loss of its hydrogen atom and the amino group takes on alkaline characteristics by acquiring an additional hydrogen atom. Such an amphoteric compound thus will have cationic and anionic properties exhibited by parts of one and the same molecule. A large number of these amino-acids have been synthesized; about twenty occur naturally as ingredients of the proteins of plants and animals; all of these, having their amine and carboxyl groups attached to the same carbon atom, are designated alpha acids and have the formula:

$$\begin{array}{c} H \\ | \\ H_2N-\mathbf{C}-COOH \\ | \\ X \end{array}$$

where **C** is the central carbon atom and X represents radicals of different complexities. In glycine (Gly), as the simplest amino-acid, X is an atom of hydrogen; in the next more complicated acid, alanine (Ala), it is $-CH_3$. In the other amino-acids that are important constituents of proteins and are found in fossils, X has the composition shown in the following table.

| AMINO-ACID | X |
|---|---|
| Serine (Ser) | $-CH_2OH$ |
| Threonine (Thr) | $-CH(CH_3)OH$ |
| Valine (Val) | $-CH(CH_3)_2$ |
| Leucine (Leu) | $-CH_2CH(CH_3)_2$ |
| Isoleucine (Ileu) | $-CH(CH_3)C_2H_5$ |
| Aspartic acid (Asp) | $-CH_2COOH$ |
| Glutamic acid (Glu) | $-CH_2CH_2COOH$ |
| Methionine (Met) | $-CH_2CH_2SCH_3$ |
| Cysteine (½ Cys) | $-CH_2SH$ |
| Cystine (Cys) | $-CH_2S-SCH_2-$ |
| Phenylalanine (Phe) | $-CH_2C_6H_5$ |
| Tyrosine (Tyr) | $-CH_2C_6H_4OH$ |
| Lysine (Lys) | $-CH_2CH_2CH_2CH_2NH_2$ |
| Hydroxylysine (Hylys) | $-CH_2CH_2CH(OH)CH_2NH_2$ |
| Arginine (Arg) | $-CH_2CH_2CH_2NHC(NH_2){=}NH$ |
| Histidine (His) | 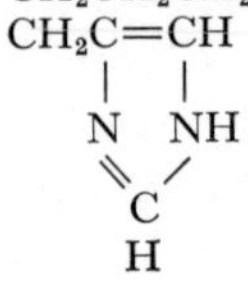 $CH_2C{=}CH$<br>│ │<br>N NH<br>⫽ /<br>C<br>H |
| Tryptophane (Try) | 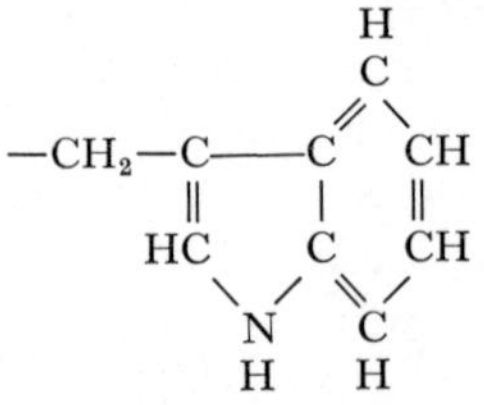 H<br>C<br>⫽ \\<br>$-CH_2-C$——C CH<br>‖ │ ‖<br>HC C CH<br>\\ / ⫽ /<br>N C<br>H H |

In addition there are the following two important acids which are pyrrolidine derivatives:

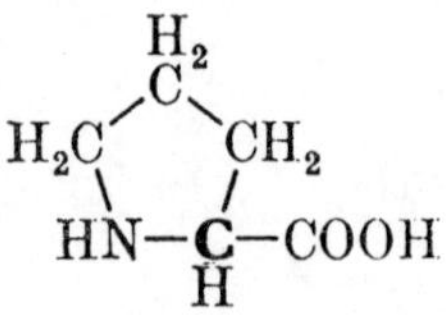

```
      H2
      C
    /   \
H2C       CH2
   \     /
   HN—C—COOH
      H
```

Proline (Pro)

```
        H
       COH
     /     \
H2C          CH2
   \        /
   HN—C—COOH
      H
```

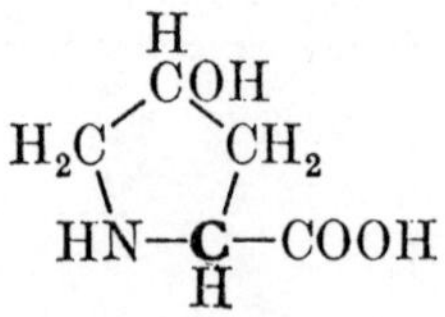

Hydroxyproline (Hypro)

The carbon atom (**C**) of each of these compounds except glycine has four different radicals attached to it and therefore is optically active; of the two isomers that thus become possible only the laevo (L) rotatory form occurs naturally as part of a protein molecule.

Molecules of these amino-acids can combine with one another through the interaction of the amino group of one molecule with the carboxyl group of another to form a dipeptide molecule, with the elimination of one molecule of water. For example, two glycine molecules, $H_2N-CH_2-CO_2H$, thus condense to form glycylglycine:

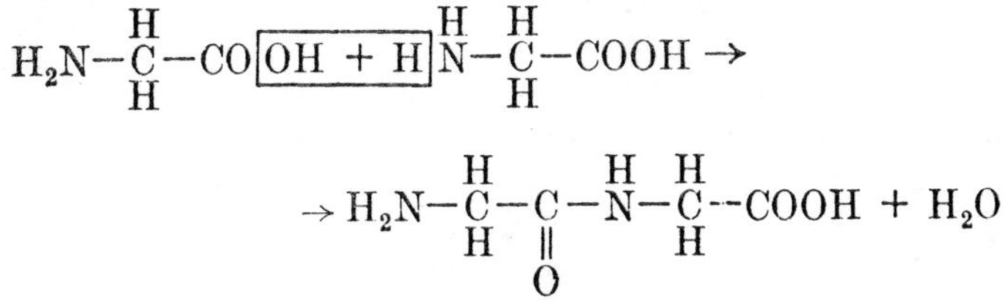

$$\begin{array}{l} \quad\;\;\; \text{H} \qquad\qquad\qquad\quad\;\; \text{H}\;\;\text{H} \\ H_2N-\text{C}-\text{CO}\boxed{\text{OH} + \text{H}}\text{N}-\text{C}-\text{COOH} \rightarrow \\ \quad\;\;\; \text{H} \qquad\qquad\qquad\qquad\;\;\; \text{H} \end{array}$$

$$\begin{array}{l} \qquad\;\; \text{H} \qquad\;\;\; \text{H}\;\;\text{H} \\ \rightarrow H_2N-\text{C}-\text{C}-\text{N}-\text{C}-\text{COOH} + H_2O \\ \qquad\;\; \text{H} \;\;\; \| \qquad\;\; \text{H} \\ \qquad\qquad\;\; \text{O} \end{array}$$

Glycyl alanine, formed by a similar reaction between a molecule of glycine and one of alanine, will have the formula:

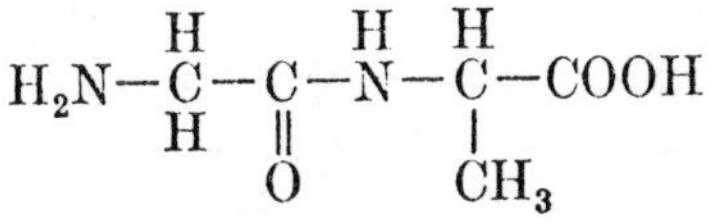

$$\begin{array}{l} \quad\;\;\; \text{H} \qquad\;\;\; \text{H}\;\;\text{H} \\ H_2N-\text{C}-\text{C}-\text{N}-\text{C}-\text{COOH} \\ \quad\;\;\; \text{H} \;\;\; \| \qquad\;\; | \\ \qquad\quad\; \text{O} \qquad CH_3 \end{array}$$

Evidently there are as many different dipeptides as there are pairs of amino-acids. Dipeptides can in turn combine with any one of the amino-acids to form a tripeptide. Longer and longer polypeptides result from the condensation of simple peptides with additional amino-acids or with one another.

In a protein molecule such chain-like polypeptides are interconnected to form a three-dimensional network whose uniqueness and almost infinitely variable complexity are the consequence of the enormous number of polypeptides of different lengths that are possible and the uncounted ways in which they can be interconnected. When such a protein molecule disintegrates, polypeptides of all lengths can be liberated and these in turn break, with the simple, 'free', acids as ultimate products. Many of the proteins of a living organism, like the amino-acids, are soluble in water or salt solutions but those forming part of the hard tissues—bones, teeth, and shells—are insoluble. Probably many of the longer polypeptide fragments also are insoluble but the shorter peptides may be nearly as soluble as the free acids and, like them, can disappear through gradual leaching. The insoluble proteinaceous residue recovered from a fossil may be the protein more or less unaltered from its original condition or it may be a mixture of this protein and large polypeptides resulting from the first steps in its disintegration. The course of this disintegration will be reflected in a progressive change in the amino-acid content of the insoluble residue from a fossil, but insolubility makes it practically impossible to decide if such a residue is in fact a three-dimensional protein or a mixture of several large polypeptides. In this monograph it will be referred to as a protein, or a proteinaceous residue, if on analysis it proves to contain most of the naturally occurring amino-acids.

The analysis of proteins has in the past been hampered by inadequate tests for their presence as well as by their extreme chemical complexity. One of the more common tests, sometimes in the past applied to fossils, has been the biuret reaction in which a blue colour is developed through interaction with a copper-containing reagent. Other substances as well as proteins are biuret-positive and nowadays more reliable tests are available. The ninhydrin reaction is currently widely used in protein analysis. Ninhydrin, $H_4C_6{-}COCOCO \cdot H_2O$, reacts quantitatively with amino groups to form very strongly coloured derivatives whose absorption of light can be made a sensitive measure of the $NH_2$-containing substance. In protein analysis this reaction is not applied to the intact protein but to its component amino-acids. A protein is analysed in terms of these acids by first splitting it down to them and then determining the amount of each in the resulting mixture. The splitting is an hydrolysis effected by heating the protein in strong acid or alkali. In most instances it is carried out by heating at about 110° C. for a day in 6*N* HCl in an atmosphere of nitrogen (to prevent oxidation).

The amino-acids are, in their chemical reactions, so similar to one another that for many years their separate identification and analysis of the mixture resulting from hydrolysis were at once complicated and unsatisfactory.

This situation has been revolutionized by the introduction of the chromatographic techniques referred to in the preceding chapter; the extensive knowledge we now have of protein structure would have been impossible without them. Numerous chromatographic procedures have been devised but, as already indicated, all involve ways to separate the amino-acids sharply from one another and to react and measure each in turn.

In chromatography as first practised, separation was effected on sheets of filter-paper but the results thus obtained could be no more than semi-quantitative. The introduction of cationic exchange resins which would take up all the amino-acids and from which they could be liberated one after another achieved quantitative results that made generally obsolete such earlier forms of chromatography. As already pointed out, the automatic amino-acid analysers now in common use yield a series of peaks of light absorption due to these successively liberated amino-acids. By measuring the peaks and comparing them with those produced when known mixtures of the pure acids are put through the apparatus, the quantity of each acid in a protein hydrolysate can be established. As customarily carried out, such liquid–column analysis requires approximately 40 millionths of a gramme (40 μg.) of each of the usual amino-acids except proline and hydroxyproline which must be present in several times this amount. This sensitivity has permitted the analysis of proteins recovered from many fossils as much as 100 million years old.

In spite of the broad range of studies thus made possible, it has for some time been apparent that still more sensitive methods of analysis would be needed to deal with the oldest fossils and the rocks that held them. Some improvement in sensitivity can be achieved by miniaturizing the conventional liquid–column instruments but the methods of gas chromatography (p. 49) seem more promising. Far smaller quantities of amino-acids can be determined in this way than by means of the ninhydrin reaction; as little as about 0·01 μM, or about a microgram (1 μg.), will furnish a measurable peak. Comparative analyses of fossil as well as pure proteins using both techniques (Akiyama and others, 1971) point to the fact that in practice samples which are twenty to fifty times smaller are all that are needed for gas chromatography.

During the application to fossils of these chromatographic methods of amino-acid analysis many problems are encountered that scarcely arise when investigating pure, undenatured proteins. With the latter, plenty of sample is usually available and the mixtures of amino-acids obtained by hydrolysis are relatively pure. So much of the original protein will have disappeared from older fossils that large samples must be taken for analysis. Their organic matter must consequently be freed from the many inorganic ions liberated when the fossil dissolves. This may not be difficult when isolating the insoluble protein fraction but it complicates analysis of the more soluble products of protein decomposition. Furthermore, some minerals may have been deposited in even the best preserved fossils and, in so far as they are soluble in hydrochloric acid, they will add to this undesirable supply of inorganic ions. Most must be eliminated to prevent contamination of the resin of the liquid–column instrument. Iron is particularly deleterious and when gas chromatography is being employed, even small amounts of this element must be removed.

It is important to see in some detail how the diverse problems encountered during the analysis of fossils can be overcome; otherwise the quality of the results obtained and future possibilities cannot be adequately appreciated. With this in mind the principal steps in a typical analysis will now be outlined.

## Analytical Procedures

Most of the protein of a fresh bone, tooth, or shell is insoluble in dilute acid and consequently what persists in a fossil should be sought in the residue from dissolution of the apatite or calcium carbonate. Soluble fragments of the original protein molecules, if present, will be found in the calcium-rich supernatant. Once a specimen has been chosen for analysis, it is important that the selected pieces be carefully inspected under a dissecting microscope to ensure the absence of mould or lichen. They should then be cleaned ultrasonically in distilled water and if necessary the surface should be removed by abrasion. There are several ways in which solution of the fossil may be effected. Ethylene diamine tetra-acetate (EDTA) can be used if the insoluble fraction alone is to be analysed but for most purposes dilute ($2N$) HCl is to be preferred. When EDTA is employed, at least 2 days must be allowed for solution to be complete and toluene or a similar bacteriostatic agent must be present to prevent growth of micro-organisms in the dissolving solution. With HCl the dissolution is prompt, there is no danger of bacterial action, and the solution can be analysed for its free amino-acids and soluble peptides. The washed sediment from the dissolution of the specimen will consist of persisting protein plus any minerals insoluble in the acid. The protein may be immediately hydrolysed to its component amino-acids by treatment for 20–24 hours at 110° C. with strong (6*N*) HCl. This should be done in a closed vessel under nitrogen to prevent oxidation. It is customary to add a crystal of phenol before the heating to inhibit further this unwanted reaction but such addition has undesirable consequences if the final analysis is by gas chromatography. The resulting solution is evaporated to dryness under reduced pressure at about 60° C. and all the remaining HCl is removed by repeated solution in doubly distilled water and re-evaporation. If the fossil is a shell and there has been little mineral deposition in it, the resultant dried mixture of amino-acids dissolved in a small (usually 1 ml.) volume of buffer may be ready for analysis by liquid–column chromatography. If, however, the fossil contained much iron-bearing or other mineral which dissolved during hydrolysis, the metallic ions thus freed should first be removed. When this is called for, the desiccated hydrolysate is dissolved in a convenient small volume of water to which is added a known amount of an unnatural amino-acid like nor-leucine. The redissolved hydrolysate is passed slowly through a short column of Amberlite 120, Dowex 50, or similar cation resin which will take up both the metallic ions and the amino-acids. After thoroughly washing the resin with water to remove chloride ions, the amino-acids are selectively eluted from it with an excess of 3–4*N* $NH_4OH$ and freed from ammonia by evaporation to dryness under reduced pressure. Dissolved finally in a small volume of buffer (for liquid–column chromatography) or of water (for gas chromatography), the purified amino-acid mixture is then ready for analysis. The presence of nor-leucine as internal standard makes unimportant any loss that may have occurred as long as it applies equally to all amino-acids. The performance of the analytical column in both liquid and gas chromatographs alters slowly with use; as a consequence standardizing runs on known amino-acid mixtures including nor-leucine should be frequent and, when purifying columns are employed, the standard mixture should first be passed through them.

The quantity of amino-acids in the final hydrolysate varies widely depending on the age and preservation of a fossil. Approximate knowledge of their concentration in the final product is therefore needed in order to choose a volume of sample giving optimal sized peaks in the chromatogram. This can be determined by carrying out a preliminary ninhydrin reaction on an aliquot (usually one-tenth) of the final sample. If the sample is so rich in amino-acids that only a small fraction of the total is used in an analysis to be made by gas chromatography, more nor-leucine must be added to ensure for it a peak of suitable size.

Additional steps in purification must be taken when determining the free amino-acids and peptide fragments in the acid solution of the fossil. They can be isolated for analysis by evaporating this solution two or three times under reduced pressure at 60° C. to remove HCl, re-solution each time being in doubly distilled water. At this point a known amount of internal standard (nor-leucine) is added to measure subsequent losses. If the fossil is a shell, the first step in purification is the elimination of calcium ions by precipitation as $CaF_2$. This is accomplished by adding a calculated slight excess of hydrofluoric acid to the aqueous solution. Calcium has sometimes been precipitated as oxalate but now that plastic flasks and centrifuge tubes are available in which to carry out operations involving hydrofluoric acid, the $CaF_2$ precipitation is much to be preferred. After centrifugation to discard $CaF_2$, traces of HF are removed by evaporating the supernatant in the rotary drier at reduced pressure. The dried deposit may be taken up in a small quantity of water and divided into two portions. When iron is absent, one can be directly analysed for the free amino-acids it contains. The other, after vacuum drying, is hydrolysed with 6*N* HCl to split its peptides into their component amino-acids. This hydrolysed portion, again vacuum dried to remove HCl, is analysed to give the free acids plus those combined as peptides. The peptide content is obtained as the difference between the two analyses.

Other problems arise when the fossil is a phosphatic tooth or bone because the phosphate would still be present as phosphoric acid after precipitation of the calcium with HF. It can be eliminated by passage of the solution, centrifuged to remove $CaF_2$, through a cathodic resin following the procedure outlined above. In carrying out this necessary removal of phosphate it must be remembered that attachment of certain amino-acids to the resin may be less than complete in the presence of such strong acid. In many instances the following method for removing calcium and phosphate has proved preferable. In applying it, the HCl solution of the fossil is vacuum dried two or three times to vaporize HCl. Some calcium phosphate remains as a precipitate when water is added. It is centrifuged out and the clear solution is made barely alkaline by the cautious addition of ammonium hydroxide; the rest of the calcium phosphate then precipitates and is discarded. Vacuum evaporation of the clear supernatant yields a solid consisting of the amino-acids and some ammonium chloride, which latter should finally be eliminated before chromatography by passage through an Araldite 120 purifying column.

These purifying resins must themselves be carefully freed of traces of protein, amino-acids, and metallic ions that may be held in them. If preliminary blank tests indicate the presence of protein, due for instance to contaminating bacteria, new resin may be heated with an excess of 2*N* NaOH which will hydrolyse protein and free the resulting amino-acids, at the same time converting the resin to its sodium form. After washing to neutrality, it must then be reconverted to the hydrogen form with strong HCl which will also liberate metallic ions that would later cause trouble. The product, washed free of all sodium and chloride ions, must finally be tested by extraction with 4*N* $NH_4OH$; if the purification has been complete, this final eluate will leave no visible residue on drying and an attempt to make volatile fluoroacetyl butyl esters for gas chromatography will be negative. Used resin can be reconverted from the ammonium to its initial hydrogen form with strong HCl.

When analysis is to be carried out by liquid–column chromatography, the amino-acid fractions as prepared and purified above can be introduced directly into the analysing apparatus. When, however, the more sensitive gas chromatography is employed, the amino-acids must next be transformed into volatile derivatives. In the Gehrke procedure (Gehrke and others, 1968), which has been satisfactorily used in fossil analysis, it was initially recommended that the thoroughly dried sample of amino-acids be converted into their butyl esters by way of methyl esters. It has now been shown that the butyl esters can be directly produced in a closed vessel at 150° C. by briefly heating the dried amino-acids with a 3*N* butyl alcohol solution of HCl:

$$H_2N-\underset{R}{\overset{H}{\underset{|}{C}}}-COOH + C_4H_9OH\left(\begin{matrix}HCl\\150^\circ\end{matrix}\right) = H_2N-\underset{R}{\overset{H}{\underset{|}{C}}}-COOC_4H_9 + H_2O$$

These butyl esters are acetylated at 100° C. with trifluoro-acetic anhydride to yield the fluoro-acetic esters:

$$H_2N-\underset{\underset{R}{|}}{\overset{H}{C}}-COOC_4H_9 + (F_3C-CO)_2O \rightarrow$$

$$\rightarrow F_3C\overset{\overset{O}{\|}}{C}-\overset{H}{N}-\underset{\underset{R}{|}}{\overset{H}{C}}-COOC_4H_9$$

Because of its limited stability the product should promptly be analysed in the gas chromatograph, preferably on the day of the synthesis. The necessary separation of these derivatives one from another depends on a proper choice of column packing and on a carefully programmed rate of heating of the column. Most column materials have been effective in separating some but not all the amino-acids and a complete analysis has involved runs on more than one column. The amino-acids actually present in fossil proteins can, however, be adequately separated by a single column using the packing recommended by Gehrke: ethylene glycol adipate impregnating 80/100 mesh heat-treated and acid-washed Chromosorb G (0·325 w/w per cent EGA).

A typical gas chromatogram of a mixture of these amino-acid derivatives is shown in *Fig.* 42. Chromatograms of similar appearance are furnished by liquid–column analysis but the amounts of amino-acid to which peaks correspond are differently computed. In liquid–column chromatography the amount of light absorbed by each coloured ninhydrin derivative is measured; the heights of the peaks therefore are logarithmic functions. There are several ways in which concentrations can be calculated from the recorded peaks; all involve estimating the areas under curves that would be obtained by transforming the observed semilogarithmic curves into their linear equivalents. The gas chromatograph, in contrast, measures concentration directly and accordingly the area under a peak of *Fig.* 42 is proportional to the quantity of the responsible amino-acid. Calculations are thus much simpler; they are readily automated and made more precise by the use of relatively inexpensive mechanical or electronic integrators that record peak position (retention time) and area.

The great sensitivity of these chromatographic methods ensures that the small quantities of proteinaceous substances preserved in ancient fossils can be quantitatively analysed (Akiyama and others, 1971). This very sensitivity, however, makes necessary special precautions to yield results representative of the entire fossil and to prevent the introduction of contamination from proteins extraneous to the specimen. In a large fossil the amount of preserved organic matter may vary widely from point to point and it is therefore better to pulverize much more than is needed

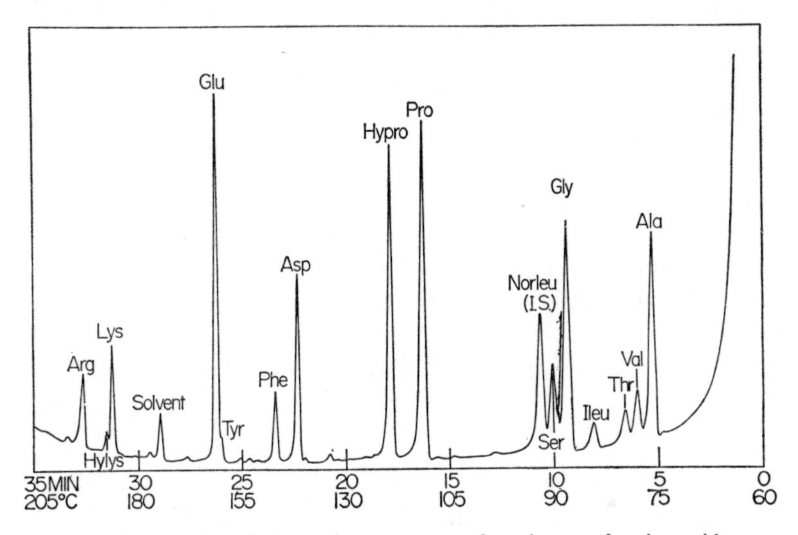


*Fig.* 42.—A typical gas chromatogram of a mixture of amino-acids converted to their volatile butyl trifluoroacetic esters. Ordinates are measures of the current through the flame; abscissae are retention times (in minutes) of the several esters and column temperatures. The peak labelled Norleu is due to 0·1 μM of nor-leucine added to the mixture as standard.

and analyse an aliquot. Contaminating proteinaceous material can come from a number of sources: it may be present in the fossil itself or its enveloping matrix, it may be introduced during collection and subsequent handling, or it may come from reagents employed in specimen preparation and analysis.

The danger of environmental contamination is minimized when the fossil is removed from a dense matrix at the time of analysis. It may usually be ignored if amino-acids are shown to be absent from this matrix. When, on the contrary, the fossil has lain exposed for some time to the weather, it is necessary to be sure that protein has not been introduced from the soil or from mould and other micro-organisms that may have been growing upon it.

These will usually manifest themselves by suspiciously high final yields of protein, and the responsible micro-organisms can themselves be seen if a bit of the residue after solution of the fossil is examined microscopically.

Specimens should in fact be examined under the dissecting microscope for adhering mould or lichen before being cleaned and washed. They can be freed of most dirt and rock fragments by sonic washing in water but often surface layers must be abraded away before final selection and weighing. During these procedures rubber gloves should be worn because sweat and minute bits of skin will contribute amino-acids to vitiate the results, especially when dealing with older specimens. Some check upon this last type of contamination is provided by the fact (Hare, 1965) that handling will make the analyses abnormally high in serine and ornithine, the latter not being a natural constituent of proteins. In earlier years it was the practice to use animal glue to reconstruct a fossil from its fragments; older museum specimens must

*Table I.*—AMINO-ACID COMPOSITION OF TYPICAL MODERN COLLAGENS (MOLE PER CENT)

| AMINO-ACID | BONE COLLAGEN* | SURGICAL GUT† | OX COLLAGEN† | RAT-TAIL TENDON† | HORSE FEMUR† | CAMEL JAW† | BOBCAT RADIUS† | BISON RIB† |
|---|---|---|---|---|---|---|---|---|
| Hypro | 10·08 | 7·96 | 10·25 | 9·42 | 8·87 | 9·00 | 9·70 | 9·37 |
| Asp | 4·98 | 4·47 | 4·25 | 4·50 | 4·53 | 4·87 | 4·96 | 4·61 |
| Thr | 1·97 | 1·96 | 1·86 | 1·99 | 1·92 | 2·33 | 1·96 | 2·00 |
| Ser | 3·78 | 3·30 | 3·48 | 4·30 | 3·60 | 3·74 | 3·51 | 3·52 |
| Glu | 7·58 | 7·21 | 6·38 | 7·10 | 7·32 | 7·55 | 7·57 | 7·69 |
| Pro | 11·87 | 11·56 | 11·81 | 12·20 | 11·85 | 11·48 | 11·91 | 11·56 |
| Gly | 31·45 | 35·79 | 31·69 | 33·10 | 32·23 | 31·88 | 31·97 | 32·32 |
| Ala | 10·97 | 10·88 | 10·79 | 10·70 | 11·22 | 10·28 | 10·55 | 11·31 |
| Val | 2·12 | 2·12 | 2·44 | 2·29 | 2·74 | 2·66 | 2·44 | 2·34 |
| Met | 0·51 | 0·38 | 0·50 | 0·84 | 0·49 | 0·67 | 0·70 | 0·64 |
| Ileu | 1·23 | 1·17 | 1·34 | 0·96 | 1·20 | 1·08 | 1·02 | 1·24 |
| Leu | 2·79 | 2·51 | 3·01 | 2·36 | 2·84 | 3·06 | 2·68 | 2·84 |
| Tyr | 0·29 | 0·79 | 0·72 | 0·39 | 0·44 | 0·59 | 0·53 | 0·47 |
| Phe | 1·63 | 1·36 | 1·93 | 1·19 | 1·49 | 1·70 | 1·35 | 1·49 |
| Hylys | 0·64 | 0·65 | 0·82 | 0·99 | 0·46 | 0·50 | 0·46 | 050 |
| His | 0·58 | 0·54 | 0·61 | 0·41 | 0·65 | 0·50 | 0·50 | 0·45 |
| Lys | 2·62 | 2·38 | 3·03 | 2·69 | 2·87 | 3·09 | 2·98 | 2·72 |
| Arg | 4·90 | 4·96 | 5·03 | 5·00 | 5·34 | 5·06 | 5·20 | 4·93 |
| Total (mg. per g.) | — | — | 223·2 | — | 184·3 | 258·1 | 177·2 | 227·4 |

*Table I—continued*

| Amino-acid | Black Bear Meta-podial† | Coyote Radius† | Arma-dillo Ulna† | Cow Whole Tooth† | Whale-skin Gela-tin‡ | Croco-dile-skin Gela-tin‡ | Shark-skin Gela-tin‡ | Cod-bone Gela-tin‡ |
|---|---|---|---|---|---|---|---|---|
| Hypro | 9·32 | 9·31 | 7·76 | 7·66 | 8·91 | 9·28 | 7·85 | 5·86 |
| Asp | 5·11 | 4·78 | 4·78 | 5·77 | 4·63 | 4·55 | 4·26 | 5·18 |
| Thr | 2·16 | 2·18 | 2·94 | 1·70 | 2·40 | 2·21 | 2·58 | 2·38 |
| Ser | 3·62 | 3·38 | 4·06 | 3·62 | 4·10 | 4·21 | 4·46 | 6·98 |
| Glu | 7·56 | 7·49 | 7·38 | 7·26 | 6·96 | 7·28 | 6·58 | 7·20 |
| Pro | 12·04 | 12·84 | 11·67 | 11·12 | 12·82 | 12·78 | 11·35 | 9·93 |
| Gly | 32·17 | 33·17 | 33·13 | 33·00 | 32·52 | 32·40 | 33·32 | 34·85 |
| Ala | 10·61 | 10·27 | 11·25 | 10·12 | 11·05 | 11·39 | 11·89 | 10·66 |
| Val | 2·41 | 2·29 | 2·62 | 1·94 | 2·06 | 1·55 | 2·19 | 1·83 |
| Met | 0·71 | 0·64 | 0·58 | 1·51 | 0·47 | 0·65 | 1·00 | 1·40 |
| Ileu | 1·05 | 0·78 | 1·02 | 1·43 | 1·10 | 1·13 | 1·94 | 1·16 |
| Leu | 2·66 | 2·38 | 2·91 | 3·18 | 2·48 | 2·01 | 2·39 | 2·30 |
| Tyr | 0·48 | 0·43 | 0·38 | 0·46 | 0·36 | 0·33 | 0·14 | 0·33 |
| Phe | 1·47 | 1·41 | 1·52 | 1·77 | 1·30 | 1·78 | 1·39 | 1·14 |
| Hylys | 0·49 | 0·56 | 0·54 | 1·14 | 0·58 | 0·49 | 0·47 | 0·82 |
| His | 0·50 | 0·47 | 0·78 | 0·75 | 0·57 | 0·47 | 0·76 | 0·74 |
| Lys | 2·66 | 2·90 | 1·93 | 2·37 | 2·59 | 2·54 | 2·43 | 2·33 |
| Arg | 4·99 | 4·71 | 4·77 | 5·22 | 5·01 | 4·96 | 5·03 | 4·85 |
| Total (mg. per g.) | 251·3 | 204·8 | 174·6 | 177·3 | — | — | — | — |

* Eastoe (1955). † Analyses from this laboratory: unpublished data; Ho (1965, 1966, 1967a). ‡ Tristram and Smith (1963).

therefore be examined closely to be certain that portions chosen for analysis are free of glue. The plastic adhesives now in use do not introduce this difficulty. The commercially available reagents, especially hydrochloric acid and the resins, sometimes contain amino-acids; and the organic compounds used in making buffers not infrequently are found supporting bacterial growth. Distilled water is another source of contamination since micro-organisms will grow in water from large stills and after standing for some time in glass containers. Fresh doubly distilled water must be used in all steps in the analysis of fossils and to rinse the glassware which should previously have been passed through chromic acid cleaning mixture. Unless unceasing vigilance is exercised by all persons in a laboratory devoted to this work with fossils, experience has shown that fallacious results due to one or another of these causes are inevitable.

*Table II.*—Amino-acid Composition of Rancho La Brea Pleistocene Bones (Mole per cent)*

| Amino-acid | Lion-like Cats | | | | Bison | | Camel | Horse | Ungulate | Bear | Wolf | Gopher |
|---|---|---|---|---|---|---|---|---|---|---|---|---|
| | V64148 | V64149 | V656 | V6522 | V657 | V6526 | V655 | V6515 | V63244 | V6521 | V63247 | V64184 |
| Hypro | 8·62 | 7·83 | 8·51 | 9·39 | 9·85 | 9·37 | 8·65 | 9·83 | 9·94 | 9·33 | 9·21 | 11·04 |
| Asp | 4·49 | 5·37 | 4·58 | 4·90 | 4·72 | 4·61 | 4·46 | 4·42 | 4·59 | 4·22 | 4·73 | 4·85 |
| Thr | 2·23 | 2·38 | 2·25 | 2·11 | 1·86 | 2·00 | 2·36 | 2·14 | 2·42 | 2·13 | 2·17 | 2·18 |
| Ser | 3·28 | 3·78 | 3·31 | 3·26 | 3·25 | 3·52 | 3·66 | 3·34 | 4·40 | 3·42 | 3·48 | 3·82 |
| Glu | 7·57 | 8·04 | 7·46 | 7·01 | 7·19 | 7·69 | 7·57 | 7·47 | 6·83 | 7·05 | 7·37 | 7·77 |
| Pro | 11·84 | 12·55 | 12·52 | 12·46 | 11·84 | 11·56 | 12·10 | 12·21 | 11·63 | 12·06 | 11·41 | 10·41 |
| Gly | 34·55 | 31·67 | 33·97 | 32·57 | 33·64 | 32·32 | 32·92 | 31·49 | 33·91 | 33·85 | 34·13 | 33·47 |
| Ala | 11·55 | 13·89 | 11·02 | 13·01 | 11·84 | 11·31 | 12·52 | 12·60 | 11·16 | 12·39 | 11·67 | 11·02 |
| Val | 2·27 | 1·96 | 2·15 | 2·35 | 2·17 | 2·34 | 2·67 | 2·36 | 2·19 | 2·31 | 2·31 | 2·30 |
| Met | 0·50 | 0·40 | 0·50 | 0·51 | 0·38 | 0·64 | 0·22 | 0·37 | 0·04 | 0·41 | 0·29 | 0·04 |
| Ileu | 0·96 | 0·83 | 1·20 | 0·98 | 1·09 | 1·24 | 0·94 | 1·04 | 1·16 | 0·92 | 1·19 | 1·06 |
| Leu | 2·32 | 2·57 | 2·39 | 2·28 | 2·43 | 2·84 | 2·48 | 2·47 | 2·76 | 2·11 | 2·40 | 2·58 |
| Tyr | 0·03 | 0·01 | 0·05 | 0·08 | 0·04 | 0·47 | 0·04 | 0·02 | 0·04 | 0·44 | 0·01 | 0·05 |
| Phe | 1·11 | 1·24 | 1·27 | 1·18 | 1·22 | 1·49 | 1·33 | 1·18 | 1·35 | 1·13 | 1·12 | 1·27 |
| Hylys | 0·44 | 0·44 | 0·31 | 0·36 | 0·41 | 0·50 | 0·48 | 0·45 | 0·36 | 0·28 | 0·51 | 0·99 |
| Orn | 0·15 | — | 0·09 | 0·11 | 0·20 | — | — | — | — | 0·10 | — | — |
| His | 0·29 | 0·46 | 0·46 | 0·30 | 0·22 | 0·45 | 0·22 | 0·26 | 0·36 | 0·25 | 0·30 | 0·46 |
| Lys | 2·75 | 2·88 | 3·11 | 2·77 | 2·86 | 2·72 | 2·39 | 2·73 | 2·15 | 3·01 | 2·81 | 2·00 |
| Arg | 5·21 | 3·69 | 4·90 | 4·92 | 4·98 | 4·93 | 4·99 | 5·61 | 4·69 | 4·66 | 4·93 | 4·70 |
| Total (mg. per g.) | 16·1 | 62·8 | 40·0 | 143·3 | 181·9 | 227·4 | 104·1 | 118·8 | 9·3 | 38·9 | 112·5 | 6·8 |

* Data from this laboratory: Ho (1965, 1966, 1967a).

## THE PROTEINS OF VERTEBRATE FOSSILS

In discussing the results obtained by applying these techniques of analysis to a variety of fossils, it is natural to begin with the most recent and move back from them to specimens of greater age, giving initial consideration to mummies and prehistoric human and animal remains. Many years ago there were extensive investigations with the optical microscope of pathological details in both the hard and soft tissues of human mummies and in the bones and teeth of fossil animals (reviewed in Moodie, 1923); on the chemical side hundreds of mummies have been blood-typed (Boyd and Boyd, 1937; Candela, 1939; Thieme, Otten, and Sutton, 1956). These studies and other reports in the literature (Edinger, 1929; Hecht, 1933; Brenner, 1939; Voigt, 1949; Baud and Morgenthaler, 1952; Ascenzi, 1955; Ezra and Cook, 1957; Schmidt and Sellmann, 1966; Lewin, 1967; Regöly-Mérei, 1967; Race and others, 1968) show that both structure and composition are preserved well enough to warrant a more thorough-going application to them of the analytical procedures discussed above.

As a first step, analyses have been made of groups of Pleistocene bones and teeth in which a major portion of the original protein has been preserved intact. The electron microscope reveals (p. 28) in these specimens recovered from the tar pits of Rancho La Brea in Los Angeles the characteristic 640 Å striations of collagen (*Figs.* 26 and 27). They are the remains of animals, many of which presumably died on falling into the pits and which accordingly have been preserved without the extensive decomposition suffered by the carcasses of most animals dying in the open. Fresh bone contains cellular proteins and a certain amount of glucoprotein which contributes glucosamine to an analysis, but most protein in even a fresh bone or tooth is collagen. As *Table I* shows, it has much the same composition no matter what the source. The same amino-acids are present in very similar amounts, with an unusually large amount of hydroxyproline as its distinguishing characteristic. Only for the codfish bone in the table is this amino-acid definitely in short supply. The material insoluble in dilute HCl obtained from many Pleistocene bones from the Rancho La Brea tar pits gave analyses (*Table II*) that do not depart substantially from one another or from the collagens of *Table I*. Similar analyses were provided by the dentines of several teeth (*Table III*) and by tooth-bearing jaws (*Table IV*) from the pits. Gelatin has been isolated from the antlers of late Pleistocene deer (Sinex and Faris, 1959). In view of such analytical results and the electron microscopic finding of a fine structure typical of collagen, one cannot seriously doubt that this protein has persisted with little if any alteration since Pleistocene times.

*Table III.*—AMINO-ACID COMPOSITION OF RANCHO LA BREA DENTINAL PROTEINS (MOLE PER CENT)*

| AMINO-ACID | GROUND SLOTH TOOTH V648 | LION PREMOLAR V63220 | HORSE PREMOLAR V63217 | BISON TOOTH V63218 |
|---|---|---|---|---|
| Hypro | 7·95 | 9·27 | 9·30 | 8·20 |
| Asp | 4·73 | 4·98 | 4·42 | 5·06 |
| Thr | 2·65 | 2·49 | 2·05 | 1·82 |
| Ser | 4·29 | 4·08 | 3·21 | 3·33 |
| Glu | 7·78 | 7·79 | 6·94 | 7·39 |
| Pro | 11·60 | 11·02 | 11·90 | 10·23 |
| Gly | 31·34 | 32·59 | 33·45 | 31·50 |
| Ala | 11·60 | 11·77 | 12·49 | 12·46 |
| Val | 2·82 | 2·31 | 2·58 | 2·62 |
| Met | 0·37 | 0·54 | 0·27 | 0·82 |
| Ileu | 1·69 | 1·21 | 1·19 | 1·04 |
| Leu | 3·59 | 2·90 | 2·47 | 2·91 |
| Tyr | 0·38 | 0·25 | 0·01 | 0·03 |
| Phe | 1·57 | 1·52 | 1·23 | 1·63 |
| Hylys | 0·96 | 0·30 | 0·88 | 1·82 |
| Orn | 0·33 | — | 0·27 | — |
| His | 0·25 | 0·54 | 0·28 | 0·85 |
| Lys | 1·47 | 1·72 | 1·98 | 2·58 |
| Arg | 4·95 | 4·71 | 5·33 | 5·69 |
| Total (mg. per g.) | 0·7 | 1·4 | 14·3 | 114·8 |

* Data from this laboratory: Ho (1965, 1967a).

Though the amino-acid composition is always that of unaltered collagen, the quantity of protein preserved in these fossils varies greatly from specimen to specimen. Thus the fossil bison bone V6526 (*Table II*) contains as large an insoluble protein fraction as the Recent bison rib of *Table I*, while in the gopher bone V64184 (*Table II*) not more than 5 per cent of the probable original content has survived. Still less remains in some teeth

(*Table III*). What is especially noteworthy is the fact that, however much of the original protein may have been destroyed, the remainder still has the composition of collagen.

A thorough study of the fate of the collagen of a bone or tooth should include a search for the molecular fragments resulting from its decay. These will be amino-acids and peptides which may remain lodged in the fossil of their origin or may have diffused out into the enclosing rock. They have not yet been sought in Rancho La Brea specimens. This should be an instructive investigation since diffusion into the tar must be minimal and therefore the total amino-acid content of a bone would depend on its condition when entrapped by the tar, that is, whether it came from a rotted carcass or a live animal. On this basis it may be concluded that the bison (V6526) referred to above was probably alive when it fell into the pit whereas the gopher (V64184) was already a carcass.

*Table IV*.—AMINO-ACID COMPOSITION OF RANCHO LA BREA FOSSIL JAWS (MOLE PER CENT)*

| AMINO-ACID | DIRE WOLF V6511 | COYOTE V6512 | LION-LIKE CAT V6513 | CAMEL V659 |
|---|---|---|---|---|
| Hypro | 8·91 | 9·34 | 9·66 | 9·49 |
| Asp | 4·43 | 4·53 | 4·90 | 4·75 |
| Thr | 2·20 | 2·21 | 2·29 | 1·89 |
| Ser | 3·49 | 3·28 | 3·35 | 2·77 |
| Glu | 7·78 | 7·26 | 7·52 | 7·27 |
| Pro | 12·00 | 11·88 | 11·34 | 11·70 |
| Gly | 34·48 | 35·04 | 33·91 | 34·52 |
| Ala | 11·47 | 11·70 | 11·10 | 12·56 |
| Val | 2·34 | 2·24 | 2·37 | 2·30 |
| Met | 0·46 | 0·48 | 0·55 | 0·40 |
| Ileu | 1·12 | 1·04 | 0·72 | 1·16 |
| Leu | 2·30 | 2·30 | 2·45 | 2·36 |
| Tyr | 0·05 | 0·10 | 0·03 | 0·08 |
| Phe | 1·20 | 0·92 | 1·15 | 1·19 |
| Hylys | 0·44 | 0·61 | 0·35 | 0·51 |
| Orn | 0·03 | 0·06 | 0·08 | — |
| His | 0·10 | 0·21 | 0·37 | 0·16 |
| Lys | 2·77 | 2·24 | 2·63 | 2·31 |
| Arg | 4·56 | 4·62 | 5·30 | 4·65 |
| Total (mg. per g.) | 101·2 | 141·3 | 146·6 | 112·4 |

* Analyses from this laboratory: unpublished data; Ho (1967a).

In fresh bones the collagen can be freed from other proteins that may accompany it by procedures involving saline extraction. If EDTA is used to demineralize a bone, non-collagenous material can be extracted with a phosphate or other neutral buffer. Using dilute HCl as demineralizing agent, extraction techniques (Partridge, 1948) have involved treatment with 10 per cent $CaCl_2$ at *p*H 9·0. These procedures, applied to the insoluble fractions from a few Rancho La Brea bones, have given results like those of *Table V*. In the case of the fossil horse bone V642, though its

*Table V*.—AMINO-ACID CONTENT OF EXTRACTED PLEISTOCENE COLLAGENS (MOLE PER CENT)*

| AMINO-ACID | HORSE, V642 | | | GROUND SLOTH, V658 | |
|---|---|---|---|---|---|
| | Acid-insoluble | $CaCl_2$-treated | Buffered | $CaCl_2$-treated | Buffered |
| Hypro | 7·72 | 9·79 | 9·53 | 8·09 | 8·41 |
| Asp | 5·22 | 4·92 | 4·57 | 4·72 | 4·96 |
| Thr | 1·74 | 1·92 | 1·86 | 1·88 | 2·00 |
| Ser | 3·57 | 3·72 | 3·35 | 3·21 | 3·65 |
| Glu | 6·99 | 7·91 | 7·28 | 7·63 | 8·13 |
| Pro | 12·50 | 10·83 | 11·72 | 11·28 | 11·07 |
| Gly | 33·89 | 32·01 | 32·55 | 34·69 | 33·54 |
| Ala | 11·29 | 13·02 | 12·74 | 12·86 | 12·24 |
| Val | 2·53 | 2·02 | 2·50 | 2·60 | 2·30 |
| Met | 1·32 | 0·38 | 0·19 | 0·42 | 0·42 |
| Ileu | 1·08 | 1·12 | 1·22 | 1·26 | 1·44 |
| Leu | 2·60 | 2·61 | 2·50 | 2·83 | 2·61 |
| Tyr | 0·03 | 0·08 | 0·06 | 0·03 | 0·03 |
| Phe | 1·44 | 1·23 | 1·22 | 1·24 | 1·24 |
| Hylys | 0·52 | 0·61 | 0·59 | 0·54 | 0·69 |
| His | 0·50 | 0·20 | 0·24 | 0·33 | 0·29 |
| Lys | 2·67 | 2·55 | 2·47 | 2·77 | 2·83 |
| Arg | 4·33 | 5·11 | 5·44 | 4·06 | 4·16 |
| Total (mg. per g.) | 200·0 | 101·6 | 141·3 | 89·3 | 90·3 |

* Analyses from this laboratory: Ho (1965, 1966).

total insoluble fraction has the composition of collagen, much of it is extractable. The composition of the residue is not significantly altered by the salt treatment and this suggests that a portion of the original collagen had been partially gelatinized or perhaps split into large acid-insoluble peptides. In some cases, as with this fossil horse bone, the amount extracted has depended on the treatment; with other fossils, such as V658 in the table, there has not been this dependence. Sometimes these 'extractables' have accounted for as much as 80 per cent of the insoluble fraction, in other fossils they have amounted to less than 10 per cent. From the very preliminary observations thus far made it is evident that much could be learned in this way about the steps through which collagen breaks down during fossilization.

Minor differences do exist in the amino-acid composition of the fossil collagenous residues of the tables but it is hard to know how much significance should be attached to them. In spite of the close similarity between the collagens from all living animals, there are real but small differences and it will be important to ascertain if fossil collagens also show them. The most striking of these is the reduced amount of hydroxyproline in the collagen of fish living in cold waters (Gustavson, 1953, 1956; Eastoe, 1957). In other living animals it has been found that the hydroxyproline and total imino-acid (proline plus hydroxyproline) contents of collagen seem to depend on both body and environmental temperatures; it is highest in birds and progressively lower in reptiles and fish (Leach, 1957). In a preliminary study it was used to assign body temperatures to Pleistocene animals from the Rancho La Brea pits (Ho, 1967b).

The excellence of preservation of the collagen in these Pleistocene fossils emphasizes the desirability of studying other immunologically more distinctive proteins which may have been preserved in fossilized soft tissues. The mastodons unearthed from the Arctic permafrost are a source of such tissues and so are the mummified animals sometimes encountered as fossils. It has been shown that characteristic immune reactions, such as those involved in blood-typing (Boyd and Boyd, 1937), can be obtained from human mummies. If other mummified soft tissues can be caused to react in this way, the results would be the source of much valuable information bearing on problems of evolution. Even more fruitful is likely to be an eventual application to fossil protein residues of the new techniques for establishing the sequence of amino-acids in a protein molecule and its component peptides. This will be difficult research but the chances of success with prehistoric human and Pleistocene animal material are good, and limited success with still older proteinaceous residues seems probable. It represents a new way to attack

*Table VI.*—AMINO-ACID CONTENTS OF PLEISTOCENE AND EARLY MAN FOSSIL PROTEINS (MOLE PER CENT)*

| AMINO-ACID | BONE V6721 | BONE V6731 | TUSK V6733 | BONE V6734 | TOOTH V6643 | TUSK V63456 | TUSK V64118 | BONE V63378 | BONE V63291 | BONE V63289 | BONE V63298 |
|---|---|---|---|---|---|---|---|---|---|---|---|
| Hypro | 8·37 | 8·47 | 9·25 | 9·04 | 7·93 | 4·90 | — | Trace | — | — | — |
| Asp | 4·08 | 3·87 | 3·80 | 4·02 | 3·44 | 7·92 | 8·07 | 11·93 | 12·22 | 10·84 | 10·41 |
| Thr | 1·30 | 1·25 | 1·33 | 1·30 | 1·36 | 3·02 | 4·87 | 5·78 | 4·93 | 5·53 | 4·06 |
| Ser | 2·95 | 3·34 | 3·31 | 2·91 | 3·33 | 4·92 | 5·34 | 4·35 | 5·34 | 3·99 | 4·53 |
| Glu | 7·00 | 6·74 | 6·58 | 7·08 | 7·34 | 11·39 | 10·15 | 14·32 | 13·52 | 13·79 | 16·25 |
| Pro | 11·24 | 11·22 | 10·86 | 11·75 | 9·57 | 6·75 | 3·03 | 3·64 | 3·95 | 3·97 | 3·76 |
| Gly | 38·76 | 39·14 | 39·44 | 37·54 | 40·28 | 26·90 | 15·31 | 11·53 | 15·02 | 11·35 | 10·98 |
| Ala | 11·08 | 11·28 | 11·08 | 10·90 | 13·06 | 13·29 | 9·61 | 18·04 | 18·39 | 16·80 | 19·51 |
| Val | 1·96 | 1·91 | 1·82 | 2·11 | 2·39 | 3·28 | 9·38 | 5·63 | 5·57 | 7·65 | 6·43 |
| ½ Cys | — | — | — | Trace | Trace | Trace | — | 0·36 | 0·40 | 0·66 | — |
| Met | 0·41 | 0·41 | 0·44 | 0·36 | 0·03 | Trace | Trace | 0·75 | — | 0·54 | 0·61 |
| Ileu | 0·95 | 0·73 | 0·75 | 1·10 | 0·86 | 1·97 | 6·88 | 2·88 | 2·78 | 4·15 | 3·26 |
| Leu | 2·34 | 2·29 | 2·21 | 2·49 | 2·44 | 6·01 | 15·96 | 11·33 | 7·45 | 10·01 | 9·11 |
| Tyr | 0·14 | 0·29 | 0·28 | 0·21 | 0·38 | 0·14 | 0·47 | 1·07 | — | 0·76 | 0·23 |
| Phe | 1·25 | 1·24 | 1·00 | 1·28 | 1·02 | 0·72 | 3·26 | 1·83 | 1·96 | 2·91 | 2·52 |
| Hylys | 0·50 | 0·27 | 0·14 | 0·43 | 0·95 | Trace | — | Trace | — | Trace | — |
| His | 0·21 | 0·39 | 0·35 | 0·26 | 0·10 | 1·01 | 0·83 | 0·46 | 0·30 | 0·46 | 0·73 |
| Lys | 2·58 | 2·54 | 2·73 | 2·43 | 1·72 | 3·32 | 3·38 | 2·81 | 3·52 | 3·06 | 4·26 |
| Arg | 4·89 | 4·63 | 4·65 | 4·82 | 3·82 | 4·47 | 3·44 | 3·27 | 4·65 | 3·50 | 3·75 |
| Total (mg. per g.) | 38·3 | 144·5 | 158·8 | 7·9 | 9·2 | 0·06 | 0·02 | 0·22 | 0·57 | 0·64 | 0·09 |

* Analyses from this laboratory: unpublished data.

problems of human prehistory and palaeontology at the point where these two sciences merge.

In turning to other Pleistocene bones and teeth, few have yet been found to be as well preserved as those from the Rancho La Brea pits. Much of the early analytical work of this laboratory was carried out on specimens collected from nearby Pleistocene sites where the fossils are plentiful but the matrix is poorly consolidated rock readily permeated by ground water. The analytical results obtained have been important mainly as indicating what to expect with specimens from such an unfavourable environment. Most of the bones studied have contained more than enough residual protein for analysis though much less than most Rancho La Brea bones. Complete data, covering the free acids and peptides as well as the insoluble protein fraction, have not yet been collected, but representative results on the residual protein fraction are given in *Table VI*. Sometimes, as with the first five fossils of the table, the analyses approach that of collagen; more frequently, though the usual amino-acids are present, their relative amounts are very different. In such fossils glycine, hydroxyproline, and proline have lost their pre-eminence at the same time that aspartic and glutamic acids and leucine have increased in importance and alanine has overreached glycine. Other fossils, like the tusk V63456 of the table, have amino-acid ratios indicating that their insoluble residues are a mixture of collagen and this different proteinaceous material.

Among fossils older than the Pleistocene, non-collagenous residues are of even more frequent occurrence. Two possible explanations for their presence immediately suggest themselves. One is that the original tissue proteins were replaced by alien proteins, perhaps shortly after the death of the animal. The other is that there has been a slow degradation of the original collagen with the passage of time, leaving behind a small, more resistant residue. Laboratory experimentation is possible which can help in deciding between these alternatives.

Most animals when they die under natural conditions are either eaten by other animals, large and small, or their carcasses rot, largely through the action of micro-organisms of various sorts. It is to be expected that in the fossilized remains of such carcasses the original proteins will have been more or less completely replaced by those of the attacking organisms and products of their vital activities. Analyses do not seem to have been made of putrefied matter but they are available for certain bacteria (Vallentyne, 1964; Akiyama, Matter, and Wyckoff, 1970). As examples, the amino-acid contents of washed, dried, and defatted masses of *E. coli* and *B. megatherium* are shown in *Table VII*. They are unlike that of collagen but closely resemble one another

*Table VII*.—THE AMINO-ACID COMPOSITION OF TWO BACTERIAL PROTEINS (MOLE PER CENT)*

| AMINO-ACID | *B. megatherium* | *E. coli* AVERAGE |
|---|---|---|
| Hypro | 0·1 | Trace |
| Asp | 11·3 | 11·0 |
| Thr | 6·3 | 6·8 |
| Ser | 5·4 | 5·2 |
| Glu | 13·0 | 11·1 |
| Pro | 4·3 | 4·7 |
| Gly | 10·2 | 10·8 |
| Ala | 15·8 | 13·0 |
| Val | 8·5 | 6·8 |
| ½ Cys | — | 0·2 |
| Met | 1·5 | 1·4 |
| Ileu | 5·2 | 4·4 |
| Leu | 10·4 | 8·8 |
| Tyr | — | 2·4 |
| Phe | 3·3 | 3·8 |
| Orn | 0·1 | — |
| His | — | 2·1 |
| Lys | 4·2 | 4·7 |
| Arg | 0·5 | 3·5 |

* Analyses from this laboratory: unpublished data.

and also that of the insoluble residue of such a fossil as V63289 of *Table VI*. Further work is obviously required, but the similarity in composition is consistent with the tentative assertion that this bone came from a carcass that had rotted before fossilization.

An idea of what probably happens to collagen as it ages can be gained through laboratory studies of its stability at elevated temperatures. Since primitive times it has been known that when boiled with water, collagen is solubilized into gelatin without, as *Table I* shows, a significant change in amino-acid composition. Information about its fate in fossils can be gained by heating moistened and dried samples in sealed tubes. Damp tendon, as absorbable sutures or after partial purification to remove most of the glucoproteins it contains, rapidly darkens, shrinks, and ultimately liquefies when heated to about 110° C. Most of the gelatinized product is very soluble in water; an insoluble residue, if there is one, continues to have the composition of collagen. After a preliminary drying for at least a day at 110° C., collagen behaves very differently at higher temperatures. Shrinkage and darkening occur at 120° C. and higher, but the hydrolytic decomposition that takes place when even a trace of moisture is present is arrested. A thorough investigation of the thermal stability of this dried collagen is not yet complete, but preliminary results indicate that though the quantity of recoverable protein declines with increase in time and temperature, the residues continue to have the approximate composition of collagen (*Table VIII*) until

*Table VIII.*—COMPOSITIONS OF HEATED COLLAGENS (MOLE PER CENT)*

| AMINO-ACID | 122° C. | | | 132° C. |
|---|---|---|---|---|
| | Dried 8 days | Dried 14 days | Dried 50 days | Dried 7 days |
| Hypro | 8·5 | 7·7 | 4·6 | 5·7 |
| Asp | 4·5 | 4·5 | 4·0 | 3·7 |
| Thr | 2·0 | 1·5 | 0·8 | 0·7 |
| Ser | 2·7 | 1·8 | 1·0 | 0·4 |
| Glu | 6·7 | 6·6 | 6·1 | 4·5 |
| Pro | 10·6 | 11·6 | 7·7 | 8·6 |
| Gly | 38·9 | 37·6 | 32·9 | 37·5 |
| Ala | 11·6 | 12·6 | 15·9 | 18·8 |
| Val | 2·4 | 2·3 | 5·9 | 6·7 |
| Met | 0·3 | 0 | 0·1 | 0 |
| Ileu | 1·2 | 1·2 | 2·2 | 2·0 |
| Leu | 2·3 | 2·3 | 5·5 | 3·8 |
| Tyr | 0·4 | 0·6 | 1·6 | 0·9 |
| Phe | 1·2 | 1·5 | 2·6 | 2·0 |
| Orn | 0·4 | 0·9 | 2·5 | 0·9 |
| Lys | 1·7 | 2·0 | 2·5 | 1·4 |
| Arg | 4·5 | 5·3 | 3·9 | 2·5 |
| Retained (per cent) | 35·7 | 11·9 | 1·2 | 0·9 |

* Analyses from this laboratory: unpublished data.

they are nearly all destroyed. Even then the changes are not great, being mainly seen as relative increases in alanine and valine with compensatory declines in proline, hydroxyproline, and glutamic acid. They show that collagenous protein will not

*Table IX.*—Amino-acid Contents of Several Older Fossil Proteins (Mole per cent)*

| Amino-acid | Pliocene V6820 | Miocene | | | Oligocene V6835 | Eocene | | | Cretaceous V696 |
|---|---|---|---|---|---|---|---|---|---|
| | | V6816 | V64117 | V67116 | | V64113 | V64142 | V6850 | |
| Hypro | — | — | — | — | 0·4 | 7·4 | — | — | — |
| Asp | 8·6 | 8·6 | 8·4 | 11·6 | 8·3 | 5·7 | 9·0 | 14·2 | 10·0 |
| Thr | 5·2 | 3·2 | 4·0 | 4·6 | 8·3 | 2·0 | 2·5 | 6·6 | 4·8 |
| Ser | 11·7 | 14·2 | 11·8 | 7·3 | 7·5 | 4·9 | 2·7 | 7·4 | 8·0 |
| Glu | 12·8 | 13·5 | 12·0 | 11·3 | 8·2 | 8·5 | 14·9 | 11·5 | 10·2 |
| Pro | 4·2 | 3·3 | 4·8 | 5·2 | 3·3 | 12·8 | 6·1 | 4·8 | 4·1 |
| Gly | 23·8 | 21·6 | 18·7 | 17·4 | 15·3 | 26·9 | 13·5 | 15·7 | 22·5 |
| Ala | 6·1 | 5·5 | 8·1 | 9·0 | 15·3 | 8·8 | 20·4 | 9·5 | 8·3 |
| Val | 4·0 | 4·2 | 5·7 | 6·1 | 8·7 | 4·0 | 7·7 | 5·3 | 5·1 |
| ½ Cys | — | Trace | Trace | — | 4·6 | — | — | — | 0·8 |
| Met | — | 0·6 | Trace | 1·2 | Trace | 0·7 | Trace | 1·0 | 0·5 |
| Ileu | 3·9 | 3·7 | 5·0 | 3·8 | 3·3 | 2·2 | 3·3 | 3·1 | 2·4 |
| Leu | 8·0 | 7·4 | 8·2 | 8·2 | 10·2 | 5·0 | 8·7 | 6·5 | 7·4 |
| Tyr | 0·4 | 1·0 | Trace | 2·1 | 0·4 | 0·7 | 0·1 | 2·7 | 1·8 |
| Phe | Trace | 2·5 | 2·8 | 3·1 | 3·1 | 2·1 | 2·4 | 2·5 | 1·8 |
| His | 1·5 | 1·5 | 1·4 | 1·0 | — | 0·9 | 0·9 | 1·3 | 1·4 |
| Lys | 4·6 | 4·6 | 4·2 | 3·9 | 0·7 | 2·8 | 3·9 | 4·1 | 4·8 |
| Arg | 5·3 | 4·6 | 4·8 | 4·4 | — | 4·6 | 4·1 | 3·9 | 4·8 |
| Total (μg. per g.) | 11·0 | 11·0 | 7·7 | 51·2 | 25·1 | 108·9 | 41·5 | 163·0 | 20·4 |

* Analyses from this laboratory: unpublished data.

be found in fossils which have been subjected simultaneously to moisture and temperatures as high as 100° C. It can likewise be asserted with considerable confidence that the non-collagenous fossil proteins noted in *Table VI* have not arisen through a progressive alteration of the initial collagen.

The poor state of preservation of many of the fossils of *Table VI* is further attested by the small insoluble residues recovered from them. The experiments just described indicate that those which still have the composition of collagen must have remained under very dry conditions. When considering the fossils which yield bacteria-like proteins, particular attention must of course be given to the possibility that these proteins are of recent origin. In many cases this uncertainty cannot be eliminated, but it can be minimized by choosing specimens that have not been long exposed to the weather but are still firmly embedded in a rock free of amino-acids. Where contamination by recent bacteria, moulds, or lichen has been encountered in this laboratory, duplicate samples have varied so much in the amount of recovered protein that this variability in itself has provided a warning signal.

Qualitative tests for the presence of amino-acids in older fossil bones (e.g., Drozdova and Kochenov, 1960; Isaacs and others, 1963) have been reported; quantitative studies are, however, still few in number and should be considered as largely exploratory in character. Typical analyses are collected in *Table IX*. Like the less well-preserved Pleistocene specimens of *Table VI*, they range all the way from the collagen-like Eocene fish bone V64113 to the bacteria-like gar scale protein V6850 of similar age. The data of this table illustrate the important fact that though the amount of proteinaceous residue in all but the best preserved Pleistocene bones is far less than in fresh specimens, it does not decline progressively with age; many that are 150 million years old have contained about the same weight of residue as those dating from a few hundreds of thousands of years.

In the investigation of older fossils a number of basic studies are needed before an adequate interpretation can be made of the results derived from individuals. Analyses should be made of numerous specimens from the same locality and of the same kind of specimen from different localities in order to define more precisely the influence of matrix on quality of preservation and to interpret whatever changes in composition are found in related specimens of increasing age.

*Table X.*—Amino-acid Contents of Pliocene Wikieup Bones and Teeth (Mole per cent)*

| Amino-acid | Bones | | | | | | | | | Teeth | | |
|---|---|---|---|---|---|---|---|---|---|---|---|---|
| | V63233 | V63539 | V63477 | V63481 | V63495 | V63526 | V63489 | V63509 | V63486 | V68108 | V6783 | V68113 |
| Hypro | — | Trace | Trace | — | — | — | Trace | — | — | — | 7·6 | 5·6 |
| Asp | 14·6 | 8·6 | 8·3 | 10·1 | 8·0 | 8·0 | 10·6 | 10·0 | 8·2 | 11·8 | 6·4 | 8·5 |
| Thr | 4·1 | 4·3 | 2·1 | 4·1 | 4·5 | 4·2 | 4·9 | 5·1 | 4·6 | 3·9 | 2·7 | 3·2 |
| Ser | 4·9 | 8·1 | 9·4 | 9·0 | 11·1 | 7·9 | 5·1 | 3·9 | 5·8 | 3·9 | 6·8 | 7·0 |
| Glu | 13·5 | 13·4 | 11·3 | 12·2 | 13·4 | 12·8 | 15·0 | 11·9 | 14·2 | 11·4 | 9·7 | 11·5 |
| Pro | 3·8 | 4·2 | 5·4 | 5·1 | 4·9 | 5·6 | 3·3 | 3·9 | 3·7 | 4·9 | 8·8 | 5·7 |
| Gly | 21·1 | 21·5 | 18·2 | 17·7 | 17·4 | 15·4 | 14·1 | 13·9 | 12·2 | 12·1 | 22·4 | 24·1 |
| Ala | 10·8 | 11·4 | 12·2 | 10·9 | 10·2 | 12·6 | 20·4 | 19·5 | 17·4 | 18·4 | 7·6 | 11·0 |
| Val | 3·5 | 3·7 | 6·3 | 5·0 | 4·7 | 5·4 | 6·0 | 5·3 | 5·1 | 6·8 | 4·3 | 4·0 |
| ½ Cys | Trace | Trace | — | — | — | — | 0·3 | — | Trace | — | Trace | 0·2 |
| Met | 0·3 | 0·4 | 0·8 | — | Trace | Trace | 0·1 | Trace | 0·8 | 0·4 | Trace | 0·4 |
| Ileu | 2·5 | 2·4 | 3·5 | 3·2 | 3·0 | 3·6 | 3·1 | 3·4 | 3·0 | 3·6 | 2·9 | 2·3 |
| Leu | 7·9 | 8·8 | 8·4 | 8·0 | 10·0 | 8·0 | 7·2 | 11·4 | 13·3 | 10·5 | 5·9 | 6·0 |
| Tyr | 1·0 | Trace | 1·1 | Trace | Trace | Trace | Trace | 1·0 | 0·2 | 0·7 | 0·5 | Trace |
| Phe | 2·5 | 3·3 | 1·8 | 2·5 | 2·4 | 2·9 | 2·3 | 1·7 | 2·4 | 2·5 | 1·8 | 2·1 |
| Hylys | Trace | — | 0·4 | — | — | — | — | Trace | — | 0·3 | 0·4 | Trace |
| His | 1·5 | 1·3 | 1·2 | 1·8 | 1·5 | 1·7 | 0·4 | 1·1 | 1·1 | 0·9 | 1·8 | 0·8 |
| Lys | 3·7 | 3·7 | 3·8 | 4·6 | 4·1 | 5·0 | 3·2 | 3·5 | 3·3 | 4·0 | 4·1 | 3·1 |
| Arg | 4·5 | 4·7 | 5·9 | 5·9 | 4·8 | 7·0 | 3·8 | 4·5 | 4·7 | 4·4 | 6·2 | 4·3 |
| Total (µg. per g.) | 93·3 | 32·7 | 34·2 | 15·2 | 26·3 | 14·1 | 54·8 | 9·9 | 47·2 | 42·1 | 19·7 | 48·7 |

* Analyses from this laboratory: Matter, Davidson, and Wyckoff (1970).

Recently a preliminary study has been published of several fossils obtained from a single site (Matter and others, 1970). The locality chosen was one in which only mediocre preservation was anticipated but which several years before had yielded a rich content of Pliocene fauna. The specimens were freshly collected fragments of bones and teeth of camels and horses embedded in a matrix that was a consolidation of the sand and silt of an ancient lake bottom. Its friable nature, together with the presence in the bones of the same calcite that cements the sand and silt, indicates a rather thorough permeation by water during fossilization. Furthermore, a content of thorium in the bones but not in the matrix suggests a prolonged contact with water even before they were engulfed in the lake-bottom silt. In spite of these unfavourable circumstances, microscopical examination showed that the basic apatitic structure was not damaged. At least 30 per cent of the specimens contained enough proteinaceous material for amino-acid analysis by the less sensitive liquid–column chromatography, the quantity in the insoluble fraction ranging up to about 100 μg. per g. of fossil. Analyses of typical bones and teeth are given in *Table X*. All the expected amino-acids are present and notwithstanding the poor environment, some of the bones have a high glycine-to-alanine ratio approaching that of collagen. Others (e.g., V63489 and V63509) have a composition approximating that of bacteria. The analyses as a whole are those to be expected if the residues of most of the bones were the result of an incomplete rotting of the original carcasses. It will be interesting to apply to such residues as these the methods that serve to purify the collagen of fresh bone. Microscopic study gave no evidence that the fossils had been the objects of recent bacterial attack and the absence of detectable amounts of protein in the matrix supports the conclusion that, whatever its origin, the age of the analysed protein was that of the fossil itself. Unlike the bones, two of the horse teeth contained the hydroxyproline characteristic of fresh collagen. These dental residues were defective in proline and contained somewhat more leucine and glutamic acid than does fresh collagen, but their composition is evidence that teeth are preserved better than bones in this relatively unfavourable environment. It emphasizes the desirability of including teeth in the fossils chosen for analysis.

Data have begun to accumulate from several other fossil-rich localities. They point clearly to the strong influence of the type of enveloping rock on the amount of organic matter retained in a fossil and the quality of its preservation. Thus far bones embedded in a calcareous matrix have been richest, but many more analyses from numerous localities must be accumulated before reliable conclusions can be drawn.

Several years ago an investigation was begun of dinosaur fossils as a potential source of proteinaceous residues older than the Pliocene. There were several reasons for this choice. One was their plentifulness, another was the large size of the single specimen. A century ago when the western United States and Canada were still being actively explored and settled and when people were first becoming familiar with the idea of evolution, dinosaurs were collected with great enthusiasm. They continue to be unearthed and in consequence of these efforts literally tons of bones and teeth are stored in our museums. The rock formations in which they are found range all the way from soft limestones to hard siliceous sandstones and shales. The state of fossil preservation is as diverse as the matrix, and dinosaur proteins provide an especially favourable experimental basis for a thorough-going examination of the relationship between them.

When the study of dinosaur proteins was begun, the low sensitivity of available chromatographic methods required samples weighing as much as 50 g. to supply enough retained protein for analysis. By using such weights, about one-third of the specimens examined gave useful results. Many of those that could not be analysed contained enough acid-soluble extraneous minerals to interfere with subsequent procedures; their presence, rather than a lack of persisting protein, became the factor that determined whether or not a successful analysis could be carried out. In bones free of this extreme mineralization, the amount of recoverable protein was comparable with that from far younger specimens. More samples became available for analysis when it was found that the iron and other interfering ions introduced by mineralization could be removed by passage through a cation resin; nevertheless this procedure is clumsy and otherwise unsatisfactory when samples of 50 g. or more must be handled. It is one of the great advantages of gas chromatography that the far smaller samples it requires have minimized this difficulty.

Typical results from Cretaceous and Jurassic bones from different formations in North America are given in *Tables XI* and *XII*. Specimens from certain localities, such as the Dinosaur

National Monument, have been generally unrewarding while from others, such as the Niobrara chalk, they have been particularly rich. It is significant for the future study of still older material that many of the Jurassic bones have retained as much protein as the younger Cretaceous. These analyses support and

*Table XI.*—AMINO-ACID CONTENTS OF JURASSIC DINOSAUR FOSSILS (MOLE PER CENT)*

| AMINO-ACID | V67138 | V674 | V67148 | V6852 | V67150 |
|---|---|---|---|---|---|
| Hypro | 1·3 | — | — | 4·8 | — |
| Asp | 12·2 | 11·5 | 12·1 | 7·8 | 8·1 |
| Thr | 5·2 | 4·7 | 4·2 | 2·1 | 7·5 |
| Ser | 6·4 | 6·6 | 7·0 | 5·8 | 7·0 |
| Glu | 11·4 | 12·6 | 12·1 | 8·4 | 11·4 |
| Pro | 5·3 | 4·3 | 4·5 | 9·0 | 2·7 |
| Gly | 16·5 | 14·8 | 17·8 | 29·4 | 19·3 |
| Ala | 10·7 | 11·3 | 11·6 | 11·1 | 8·0 |
| Val | 6·0 | 8·4 | 5·6 | 3·4 | 5·7 |
| ½ Cys | 1·3 | — | Trace | Trace | — |
| Met | 0·9 | 1·3 | 0·9 | 0·3 | 0·7 |
| Ileu | 3·0 | 3·3 | 3·8 | 1·9 | 3·9 |
| Leu | 7·1 | 8·6 | 7·3 | 4·2 | 9·7 |
| Tyr | 1·9 | 0·9 | 1·7 | 0·6 | 2·6 |
| Phe | 2·3 | 2·6 | 2·2 | 1·8 | 3·7 |
| Hylys | 0·4 | — | Trace | 0·7 | — |
| His | 1·2 | 1·2 | 1·2 | 0·7 | 1·7 |
| Lys | 4·0 | 3·9 | 3·6 | 3·2 | 3·8 |
| Arg | 4·4 | 4·2 | 4·5 | 5·1 | 4·2 |
| Total (μg. per g.) | 25·0 | 28·8 | 17·4 | 42·2 | 10·8 |

* Analyses from this laboratory: Miller and Wyckoff (1968), Wyckoff (1969, 1971).

extend the indications obtained with the younger Pliocene and Miocene specimens that the amount of protein in old fossils declines only slowly with age.

The dinosaur proteins have contained practically all the amino-acids present in younger fossils; except for reduced amounts of the relatively less stable hydroxy and sulphur-containing acids, they are those of modern proteins. There is thus no evidence in these data that the proteins of 150 million years ago were simpler in composition than those of today. As with the younger bones already discussed, the relative amounts of amino-acids have varied greatly from bone to bone. A few, notably the Cretaceous V6869 and the Jurassic V6852, have compositions approaching that of Modern collagen. Others, such as V6851 and V67148, have their amino-acids in the proportions to be expected if they were derived from rotted animal carcasses. Most have compositions suggesting they may be mixtures of collagenous and micro-organismal proteins. It is to be noted that V6852 and V67148 came from the same fossil bed and therefore were preserved under essentially the same conditions; presumably their different amino-acid ratios reflect the different states of the carcasses when fossilization began.

Analyses have been published (Armstrong and Tarlo, 1966) of hydrolysates of a Jurassic phytosaur bone and tooth and of conodonts of Devonian age. The phytosaur analyses were similar but the surrounding rocks were richer in amino-acids than the fossils themselves; for this reason no definite interpretation can be given to the results.

*Table XII.*—AMINO-ACID CONTENTS OF CRETACEOUS REPTILIAN FOSSILS (MOLE PER CENT)*

| AMINO-ACID | V67140 | V6869 | V67146 | V6837 | V67142 | V675 | V6519 |
|---|---|---|---|---|---|---|---|
| Hypro | — | 8·1 | — | — | Trace | Trace | — |
| Asp | 8·4 | 4·3 | 13·0 | 8·4 | 12·5 | 14·5 | 11·7 |
| Thr | 3·9 | 2·0 | 5·2 | 4·1 | 5·4 | 6·0 | 2·4 |
| Ser | 8·3 | 3·7 | 4·2 | 15·1 | 6·3 | 7·5 | 8·8 |
| Glu | 13·3 | 8·1 | 14·0 | 12·8 | 13·3 | 10·3 | 12·5 |
| Pro | 4·4 | 9·1 | 3·9 | 3·1 | 5·3 | 5·5 | 4·4 |
| Gly | 18·2 | 36·4 | 12·0 | 19·4 | 16·8 | 19·1 | 22·3 |
| Ala | 11·0 | 10·6 | 15·7 | 5·8 | 12·4 | 10·0 | 8·7 |
| Val | 5·2 | 2·5 | 5·6 | 4·0 | 6·4 | 5·0 | 5·4 |
| ½ Cys | — | Trace | Trace | 0·8 | — | Trace | — |
| Met | 1·2 | 0·6 | 0·7 | 0·8 | 1·1 | 0·8 | 0·8 |
| Ileu | 3·0 | 1·8 | 3·1 | 3·0 | 3·1 | 2·7 | 2·8 |
| Leu | 8·5 | 3·3 | 10·0 | 7·4 | 7·9 | 5·9 | 6·5 |
| Tyr | 2·4 | 0·5 | 1·6 | 2·7 | 1·4 | 1·1 | 2·0 |
| Phe | 1·9 | 1·3 | 2·3 | 2·7 | — | 2·7 | 2·8 |
| Hylys | — | 0·4 | Trace | — | — | Trace | 0·8 |
| His | 1·1 | 0·5 | 1·3 | 1·2 | 1·1 | 1·4 | 1·2 |
| Lys | 4·1 | 2·3 | 3·6 | 3·9 | 3·2 | 3·9 | 3·5 |
| Arg | 5·0 | 4·5 | 3·8 | 4·5 | 3·7 | 3·8 | 3·4 |
| Total (µg. per g.) | 14·9 | 74·9 | 62·2 | 29·2 | 17·3 | 31·8 | 18·9 |

*Table XII—continued*

| Amino-acid | V67147 | V6851 | V6613 | V67154 | V67155 | V67162 | V67156 |
|---|---|---|---|---|---|---|---|
| Hypro | 0·4 | — | — | — | — | — | Trace |
| Asp | 11·6 | 12·4 | 8·8 | 10·7 | 11·1 | 11·7 | 9·5 |
| Thr | 5·9 | 5·7 | 5·4 | 5·7 | 5·5 | 5·5 | 4·5 |
| Ser | 5·8 | 4·9 | 7·7 | 4·9 | 5·8 | 5·1 | 10·2 |
| Glu | 9·8 | 11·7 | 12·6 | 15·1 | 15·1 | 13·3 | 13·1 |
| Pro | 5·4 | 4·9 | 5·0 | 3·7 | 3·9 | 4·4 | 4·8 |
| Gly | 12·0 | 11·1 | 14·5 | 10·3 | 9·8 | 10·9 | 20·2 |
| Ala | 13·8 | 14·9 | 12·1 | 17·8 | 17·1 | 16·5 | 6·4 |
| Val | 7·5 | 6·4 | 5·6 | 5·9 | 6·1 | 5·9 | 4·9 |
| ½ Cys | 0·8 | 0·2 | — | 0·2 | 0·2 | Trace | — |
| Met | 2·4 | 1·2 | Trace | 1·1 | 1·1 | 1·2 | 0·7 |
| Ileu | 2·7 | 3·5 | 3·4 | 3·5 | 3·5 | 3·3 | 3·8 |
| Leu | 5·4 | 8·5 | 8·7 | 8·9 | 8·3 | 8·1 | 7·8 |
| Tyr | 0·8 | 1·7 | 2·2 | 1·5 | 1·2 | 1·8 | 1·8 |
| Phe | 2·3 | 2·9 | 2·7 | 2·7 | 2·7 | 2·8 | 2·8 |
| Hylys | — | — | 0·6 | — | — | — | Trace |
| His | — | 1·2 | 1·6 | 0·6 | 0·8 | 1·1 | 1·3 |
| Lys | 2·9 | 3·9 | 4·3 | 3·1 | 3·1 | 3·6 | 3·6 |
| Arg | — | 4·9 | 4·9 | 4·3 | 4·2 | 4·8 | 4·7 |
| Total (µg. per g.) | 5·9 | 138·7 | 31·3 | 296·0 | 225·7 | 165·4 | 21·4 |

* Analyses from this laboratory: unpublished data; Miller and Wyckoff (1968), Wyckoff (1969, 1971).

During these early stages in the investigation of fossil vertebrate proteins attention has been concentrated on the analysis of single bones and teeth and of their enveloping rocks. This choice has of course been based on the presumption that insoluble organic matter present in the fossil but absent from its matrix must have originated where found. As one recedes in time, it will be impossible to deal only with such individual specimens. The Triassic and Permian giant amphibians provide large bones but earlier than this one must be content with less substantial remains of fish. Then the samples will often be aggregates of small bones and teeth. A few years ago they would have been beyond the reach of our analytical methods, but the great sensitivity of gas chromatography ensures that we can now undertake with confidence the examination of the most ancient vertebrate fossils.

There are a few other fossils of vertebrate origin which might be expected to contain proteins unrelated to collagen. One of these is tooth enamel. The enamel of fresh adult teeth contains less than 1 per cent of a keratin-like protein which has been difficult to analyse because the thin enamel layers of the teeth of most animals are not easily freed from bits of protein-rich underlying dentine. There have been good analyses of cow tooth enamel but even more satisfactory results could be obtained by working with the giant molars of members of the elephant family. Their enamel is as much as a centimetre in thickness. Fossil enamel protein can be sought in the teeth of such animals which, with their land- and shore-living relatives, go back through the Eocene. A beginning has been made through the analysis of the molar of a Pleistocene mammoth (*Table XIII*). Departing far from that of fossil collagen, its composition in a general way

*Table XIII.*—AMINO-ACID CONTENT OF MODERN AND FOSSIL TOOTH ENAMEL,* HAIR,† AND OTOLITHS‡ (MOLE PER CENT)

| AMINO-ACID | ENAMEL | | HAIR | | OTOLITHS | |
|---|---|---|---|---|---|---|
| | Modern Bovine | Pleistocene V6524 | Modern Elephant | Pleistocene Mammoth | Modern (average) | Fossil Miocene |
| Hypro | 1·2 | 2·5 | — | — | 3·2 | 0 |
| Asp | 11·8 | 7·8 | 7·08 | 8·51 | 16·2 | 2·2 |
| Thr | 4·5 | 3·5 | 7·24 | 6·27 | 6·0 | 0·6 |
| Ser | 12·3 | 10·5 | 11·00 | 7·02 | 6·0 | 1·1 |
| Glu | 11·4 | 11·7 | 14·20 | 17·50 | 16·8 | 21·3 |
| Pro | 6·7 | 4·9 | 7·60 | 5·59 | 5·3 | 10·9 |
| Gly | 14·9 | 20·6 | 6·50 | 4·37 | 12·6 | 5·1 |
| Ala | 9·1 | 8·1 | 6·06 | 6·94 | 9·4 | 30·3 |
| Val | 4·8 | 5·0 | 6·03 | 6·45 | 6·9 | 13·4 |
| ½ Cys | 1·1 | 0·4 | 6·80 | 7·42 | 1·9 | 10·0 |
| Met | 0·1 | 1·0 | 0·47 | 0·56 | 0·1 | — |
| Ileu | 2·9 | 3·5 | 3·68 | 4·17 | 3·0 | 1·9 |
| Leu | 6·8 | 7·5 | 7·64 | 9·35 | 7·1 | 3·4 |
| Tyr | Trace | 2·4 | 2·77 | 2·46 | 0·1 | 0 |
| Phe | 3·2 | 2·8 | 2·10 | 1·83 | 1·2 | 0 |
| His | 1·6 | 0·5 | 0·84 | 0·72 | 0·8 | — |
| Lys | 4·6 | 3·0 | 2·74 | 3·15 | 2·2 | — |
| Arg | 3·1 | 4·4 | 6·85 | 7·19 | 1·5 | — |
| Total (μg. per g.) | 600 | 18·8 | — | — | — | 810 |

* Analyses from this laboratory: Doberenz, Miller, and Wyckoff (1969).
† Analytical data from Gillespie (1970).
‡ Analytical data from Degens, Deuser, and Haedrich (1969).

resembles that of fresh bovine enamel. There are, however, real differences, for instance in the aspartic acid and glycine contents, which emphasize the need for analyses of fresh elephant enamel and of large fossil teeth, including those of the mammoths found in the Arctic permafrost. A surprising result of the present analysis has been the finding in fossil enamel of the hydroxyproline commonly taken as diagnostic for collagen; considering the thickness of mammoth enamel ground away when preparing the sample, it is scarcely possible that this amino-acid can have come from contaminating dentine.

Another protein of similar type which can reasonably be sought in a few favourable fossil specimens is the keratin of hair. Fossil hair of Pleistocene mammoths recovered from Alaskan permafrost has been analysed (Gillespie, 1970), together with fresh hair of the Indian elephant, to give the data recorded in *Table XIII*. It has been proposed that the differences between the two samples should be ascribed to decomposition of the fossil, and, to provide evidence bearing on this interpretation, additional analyses were carried out on high and low sulphur fractions.

Still another calcified proteinaceous substance is the otolith often found in the inner ears of fish. A recent study (Degens, Deuser, and Haedrich, 1969) of a number taken from 25 different kinds of fish showed the mineral in all to be aragonite and the amount of protein to range between 0·1 and 10 per cent. The average amino-acid content is given in *Table XIII* together with an analysis of a fossil otolith of Miocene age. The fresh and fossil proteins do not resemble one another in composition and neither is like the other vertebrate proteins that have been considered.

Other fossils of vertebrate origin which should contain non-collagenous proteins are the shells of avian and certain reptilian eggs. All such shells resemble one another in consisting of crystalline calcium carbonate interspersed with protein. Most are too thin and fragile to have been preserved as part of the fossil record, but this is not the case with the thick egg-shells of the large ratite birds and of dinosaurs. Analyses have been made of proteinaceous residues from both.

The best known fossil bird egg-shells are those of the Pleistocene *Aepyornis*, found in abundance in Madagascar. The eggs of these very large, ostrich-like birds were correspondingly large and, having shells several millimetres thick, provide plenty of material for analysis. Egg-shells of other probable ancestors of the ratite birds back through the Eocene have been collected in North Africa and the south of France. Analyses of the insoluble proteinaceous residues recovered from them are collected in *Table XIV*. In order better to understand the fossil data, studies

*Table XIV*.—AMINO-ACID CONTENT OF EGG-SHELL PROTEINS OF RECENT AND FOSSIL RATITE BIRDS (MOLE PER CENT)*

| AMINO-ACID | RECENT | | | PLEISTOCENE | EOCENE |
|---|---|---|---|---|---|
| | Ostrich V6897 | Cassowary V6896 | Emu V6894 | *Aepyornis*† | *Ornitholithus* V6946 |
| Hypro | 0·6 | — | — | — | — |
| Asp | 10·7 | 8·8 | 8·1 | 10·6 | 11·6 |
| Thr | 4·5 | 5·1 | 5·0 | 3·5 | 8·7 |
| Ser | 8·2 | 8·5 | 7·7 | 3·4 | 6·2 |
| Glu | 11·5 | 9·9 | 9·9 | 13·7 | 10·9 |
| Pro | 4·9 | 8·6 | 9·4 | 3·7 | 6·7 |
| Gly | 9·2 | 9·6 | 9·6 | 8·2 | 9·8 |
| Ala | 8·2 | 8·0 | 8·1 | 10·7 | 13·3 |
| Val | 4·4 | 3·7 | 4·0 | 7·4 | 7·8 |
| ½ Cys | 3·2 | 3·8 | 2·8 | — | 0·2 |
| Met | 0·8 | 0·2 | 0·3 | 0·6 | 0·8 |
| Ileu | 4·6 | 2·8 | 3·0 | 5·5 | 4·3 |
| Leu | 7·0 | 8·9 | 9·4 | 13·1 | 9·4 |
| Tyr | 4·1 | 4·2 | 4·1 | 4·1 | 1·9 |
| Phe | 4·4 | 3·9 | 3·7 | 5·8 | 3·4 |
| Orn | 0·4 | — | — | — | 0·3 |
| His | 3·2 | 2·2 | 2·6 | 0·9 | — |
| Lys | 4·9 | 3·7 | 4·0 | 4·2 | 4·2 |
| Arg | 6·1 | 8·1 | 8·5 | 6·6 | — |
| β-Ala | 0·1 | — | — | — | — |
| Total (mg. per g.) | 8·4 | 22·0 | 19·3 | 0·58 | 0·23 |

* Analyses from this laboratory: Matter, Davidson, and Wyckoff (1971).
† Averaged from V6846 and V6931.

have been made of the shells of living ratites: the ostriches of Africa, and the cassowaries and emus of Australia. The proteins of all these shells, both Recent and fossil, are of the same type, with aspartic and glutamic acids, glycine, alanine, leucine, and arginine as their dominant amino-acids. The cassowary and emu analyses are practically identical but the ostrich protein is somewhat higher in aspartic and glutamic acids and lower in proline than the others. It is hard to place a final evaluation on the minor differences between the two fossil proteins and those of the present-day ratites because, as comparison with *Table VII* indicates, these shell proteins somewhat resemble in composition those of micro-organisms. The alanine-to-glycine ratio greater than unity of the fossil proteins makes it necessary to entertain

*Table XV*.—The Amino-acid Content of Bird Egg-shell Proteins (Mole per cent)*

| Amino-acid | Chicken V68112 | Blue Heron V6878 | Blackbird V6877 | Stilt V6879 | Magpie V6880 | Average |
|---|---|---|---|---|---|---|
| Asp | 8·1 | 7·4 | 7·6 | 8·5 | 7·1 | 7·5 |
| Thr | 6·0 | 6·9 | 6·5 | 7·1 | 6·3 | 6·4 |
| Ser | 7·0 | 8·1 | 7·6 | 8·3 | 7·9 | 7·6 |
| Glu | 10·5 | 10·1 | 10·9 | 7·5 | 10·1 | 10·4 |
| Pro | 8·0 | 7·4 | 8·2 | 7·9 | 8·6 | 8·0 |
| Gly | 11·6 | 10·8 | 10·6 | 10·3 | 10·9 | 10·9 |
| Ala | 6·0 | 8·1 | 7·5 | 6·8 | 7·4 | 7·2 |
| Val | 6·6 | 6·3 | 5·0 | 6·1 | 5·5 | 5·8 |
| ½ Cys | 4·8 | 3·8 | 4·1 | 3·6 | 5·6 | 4·6 |
| Met | 2·8 | 1·3 | 1·8 | 1·6 | 1·8 | 1·9 |
| Ileu | 3·6 | 3·3 | 3·1 | 4·0 | 2·8 | 3·2 |
| Leu | 6·1 | 6·7 | 5·8 | 7·1 | 5·7 | 6·1 |
| Tyr | 3·1 | 4·6 | 4·4 | 4·0 | 4·4 | 4·1 |
| Phe | 2·1 | 2·8 | 2·9 | 4·0 | 2·7 | 2·6 |
| Hylys | 0 | 0·6 | 0·6 | 0 | 0·4 | 0·5 |
| His | 4·2 | 2·7 | 2·8 | 3·5 | 3·2 | 3·2 |
| Lys | 4·1 | 3·1 | 2·8 | 2·6 | 3·0 | 3·2 |
| Arg | 5·7 | 6·1 | 7·7 | 7·1 | 6·3 | 6·4 |
| Total (mg. per g.) | 12·8 | 14·1 | 18·7 | 16·5 | 19·1 | |

* Analyses from this laboratory: Matter, Davidson, and Wyckoff (1971).

the possibility that at one time or another they may have undergone some bacterial attack.

In making these comparisons, it has been interesting to include analyses of the shell proteins of several Modern flying birds (*Table XV*). Except for a reduced amount of glutamic acid in the stilt protein, they are essentially identical in composition and in total amount of protein per gramme of shell. They do not greatly differ from the cassowary and emu proteins. It has often been proposed that the ratite birds as a group had a common origin far back in time and are not recent adaptations to a flightless life. The present data do not reveal significant differences between their egg-shell proteins and those of flying birds but they do show that their shell proteins have not appreciably altered since Eocene times.

*Table XVI.*—AMINO-ACID CONTENT OF DINOSAUR EGG-SHELL PROTEINS (MOLE PER CENT)*

| AMINO-ACID | UTAH | | GOBI | AIX-EN-PROVENCE | | | | |
|---|---|---|---|---|---|---|---|---|
| | V6843 | V6844 | V68101 | V7034 | V7035 | V7036 | A† | B† |
| Hypro | — | — | 3·1 | — | — | — | — | — |
| Asp | 8·7 | 9·8 | 6·5 | 9·5 | 8·6 | 7·0 | 13·3 | 10·5 |
| Thr | Trace | 4·7 | 6·1 | 7·2 | 8·0 | 5·6 | 5·5 | 6·9 |
| Ser | 15·7 | 7·1 | 8·0 | 6·5 | 5·6 | 6·2 | 11·9 | 6·6 |
| Glu | 13·3 | 17·3 | 8·4 | 10·5 | 10·5 | 8·8 | 13·3 | 11·4 |
| Pro | 3·3 | 5·1 | 5·3 | 7·1 | 7·0 | 7·5 | — | 6·4 |
| Gly | 20·7 | 12·8 | 15·7 | 8·5 | 8·5 | 13·7 | 19·7 | 10·3 |
| Ala | 5·9 | 15·1 | 8·7 | 12·6 | 13·1 | 11·7 | 9·0 | 10·0 |
| Val | 4·1 | 6·5 | 8·3 | 9·3 | 9·6 | 9·5 | 5·6 | 7·0 |
| ½ Cys | 1·4 | — | — | — | — | — | — | — |
| Met | 0·7 | — | 1·9 | 1·1 | 0·5 | 2·8 | — | 0·9 |
| Ileu | 3·2 | 4·2 | 4·4 | 6·1 | 5·3 | 5·7 | 3·6 | 4·0 |
| Leu | 7·2 | 8·2 | 9·2 | 10·8 | 11·5 | 8·6 | 7·0 | 7·6 |
| Tyr | 3·3 | 2·6 | 2·4 | 2·5 | 2·5 | 2·9 | 2·6 | 3·7 |
| Phe | 2·1 | 3·0 | 3·3 | 4·8 | 5·3 | 4·6 | 2·6 | 4·0 |
| Orn | — | — | 0·6 | — | — | — | — | — |
| His | 1·5 | — | — | — | — | — | — | 1·2 |
| Lys | 3·8 | 3·5 | 1·2 | 3·3 | 4·1 | 3·0 | 5·6 | 5·5 |
| Arg | 5·1 | — | — | — | — | — | — | 3·8 |
| Total (μg. per g.) | 28·2 | 38·5 | 12·3 | 48·6 | 152·6 | 6·4 | — | — |

* Except for A and B, analyses from this laboratory: unpublished data.
† A is the brown layer, B the outer lining as analysed in Voss-Foucart, 1968.

From the point of view of structure, the egg-shells of dinosaurs closely resemble those of the ratite birds (*Figs.* 12, 13, and 15). They are of comparable thickness and like the bird shells are built up of organic membranes intercalated between layers of calcite. Specimens were collected years ago in the Gobi Desert of central Asia and are also found in great abundance (Dughi and Sirugue, 1966) near Aix-en-Provence in France. Though the skeletal remains of dinosaurs are frequently encountered in western North America, the shells of their eggs are very scarce (Jensen, 1966, 1970).

In a first analysis of a shell from Aix it was reported (Voss-Foucart, 1968) that two separate proteins could be isolated. One (A), dark in colour, was from the internal membranes, the other (B) was a white membrane covering the shell and the air pores that traverse it. Their compositions, together with those of other dinosaur shells that have been analysed, are collected in *Table XVI*. This separation of proteins was not attained during an investigation in this laboratory of shells from the same region. The great sensitivity of gas chromatography made it possible to analyse individual shell fragments, but when this was done they were found to vary by as much as a hundredfold in their content of insoluble protein. In the attempt to explain such variations it was discovered that many of the shells were contaminated. Membranes observed in the insoluble residues of some samples were revealed by microscopical examination to be masses of mould mycelium. Firmly adherent minute colonies of lichen were also found on some shell fragments and it was apparent that with these specimens the danger of recent contamination was especially great. Such contamination was not visible in the samples giving the data of the table, but these results must be regarded as tentative until much more work has been done. There are certain similarities between some of these proteins, for instance between the Gobi Desert protein V68101 and the theropod protein from Aix (V7036), but taking the problem as a whole we have to admit that we are not yet in a position to explain most of the diversities that are evident.

In order to help in the interpretations that must be made, analyses have been carried out on the shells of several Modern reptilian eggs. The eggs of some reptiles are hard-shelled, those of other groups soft-shelled; still others are imperfectly calcified. Among the reptiles alive today, the tortoise and the crocodilian shell proteins are practically identical (*Table XVII*) though the mineral of the first is aragonite and of the second the more usual calcite. Very different are those of the lizard egg-shells which, besides being in many cases incompletely calcified, are unique in their high content of proline. Disregarding the latter, it is to be noted that these reptilian proteins are somewhat like some of those found in dinosaur egg-shells but most closely resemble those of the emu and cassowary.

It has already been pointed out (p. 52) that for a complete understanding of the proteinaceous residues in fossils it is essential also to investigate them for the fragments of protein decomposition, amino-acids, and peptides, they may have retained. The analytical techniques for doing this are now at hand but there has as yet not been time to apply them to many fossil bones

*Table XVII.*—AMINO-ACID COMPOSITION OF MODERN REPTILIAN EGG-SHELL PROTEINS (MOLE PER CENT)*

| AMINO-ACID | TORTOISES | | | CROCODILIA | | LIZARD |
|---|---|---|---|---|---|---|
| | African V6891 | Galapagos V6889 | Desert V6888 | Alligator† | Crocodile V6890 | Basilisk V6885 |
| Asp | 8·4 | 8·3 | 8·5 | 9·6 | 11·0 | 6·2 |
| Thr | 7·2 | 7·3 | 6·7 | 8·8 | 9·2 | 1·9 |
| Ser | 9·9 | 7·6 | 9·6 | 6·5 | 7·5 | 5·0 |
| Glu | 8·5 | 8·6 | 9·1 | 8·6 | 8·4 | 4·8 |
| Pro | 7·0 | 8·0 | 7·8 | 7·6 | 7·4 | 23·2 |
| Gly | 6·8 | 7·7 | 8·0 | 7·1 | 6·8 | 9·3 |
| Ala | 6·2 | 5·4 | 6·1 | 5·0 | 6·1 | 9·0 |
| Val | 6·5 | 7·2 | 7·0 | 7·1 | 5·8 | 5·6 |
| ½ Cys | 5·0 | 7·6 | 6·0 | 5·5 | 4·1 | 8·3 |
| Met | 1·6 | 0·9 | 1·6 | 1·3 | 0·9 | 0·3 |
| Ileu | 4·3 | 4·8 | 4·0 | 4·2 | 4·8 | 1·4 |
| Leu | 6·9 | 5·5 | 6·2 | 6·1 | 6·3 | 3·7 |
| Tyr | 7·0 | 6·2 | 4·7 | 6·4 | 7·3 | 8·7 |
| Phe | 3·6 | 3·8 | 2·7 | 3·3 | 2·7 | 3·1 |
| Hylys | 0 | 0·4 | 0·5 | — | — | — |
| His | 1·9 | 1·5 | 2·9 | 3·6 | 3·0 | 5·2 |
| Lys | 4·5 | 4·2 | 4·6 | 4·7 | 3·5 | 0·9 |
| Arg | 4·7 | 4·9 | 4·1 | 5·4 | 5·2 | 3·4 |
| Total (mg. per g.) | 2·65 | 4·34 | 2·93 | 15·1 | 10·8 | 42·8 |

* Analyses from this laboratory.
† Average from samples V6893 and V6898.

and teeth. In a few exploratory analyses the free amino-acids have been compared with the proteinaceous, insoluble residues. Typical data are given in *Table XVIII*. In the two bones shown here, the soluble fraction is smaller than the insoluble but in other fossils the reverse has been true. These variations presumably are expressions of the different environmental conditions to which the fossils have been subjected. Usually the composition of the soluble fraction has been quite unlike that of the proteinaceous residue but this is to be expected since, though the total amount of free acids in a fossil depends on its exposure to percolating water, the acids that are found will depend on their individual stabilities.

*Table XVIII.*—TYPICAL FREE AND COMBINED AMINO-ACID ANALYSES OF FOSSIL BONES (MOLE PER CENT)*

| AMINO-ACID | V63295 | | V63562 | |
|---|---|---|---|---|
| | Soluble | Insoluble | Soluble | Insoluble |
| Hypro | 0 | 0 | 1·7 | 0·4 |
| Asp | 1·5 | 7·4 | 2·4 | 13·3 |
| Thr | 1·2 | 3·8 | 5·2 | 7·4 |
| Ser | 4·3 | 6·2 | 4·3 | 5·7 |
| Glu | 4·2 | 11·6 | 5·7 | 7·0 |
| Pro | 5·6 | 4·1 | 4·1 | 3·9 |
| Gly | 16·8 | 15·2 | 26·5 | 15·8 |
| Ala | 15·5 | 13·6 | 10·9 | 17·8 |
| Val | 6·1 | 7·2 | 5·7 | 9·3 |
| Met | 4·2 | 0·6 | 1·4 | 0·3 |
| Ileu | 2·0 | 4·1 | 2·4 | 4·0 |
| Allo-ileu | 0·9 | 0 | 0 | 0 |
| Leu | 5·3 | 7·4 | 2·8 | 7·3 |
| Tyr | 7·7 | 2·5 | 0 | 1·0 |
| Phe | 0 | 3·1 | 1·4 | 3·1 |
| Orn | 9·2 | 3·8 | 0 | 0 |
| Lys | 2·0 | 6·4 | 2·1 | 1·5 |
| Arg | 0 | 0 | 0 | 0 |
| Try | 2·0 | 0 | 0 | 1·0 |
| β-Ala | 6·0 | 3·7 | 12·8† | 1·2 |
| γ-ABA | 5·5 | 2·5 | 10·8† | 0 |
| Total (μg. per g.) | 5·5 | 13·5 | 5·0 | 22·4 |

* Analyses from this laboratory: unpublished data.
† Peaks having the retention times of these amino-acids.

Though they are in fact very stable substances, the amino-acids do not persist unchanged for geologically important periods of time. Some slowly decompose, others transform into other amino-acids, many of which do not form part of present-day proteins. In analysing the free amino-acids of a fossil one must therefore expect to find not only those initially present in the protein but also others that have arisen from these transformations.

The most conspicuous are allo-isoleucine, β-alanine, γ-aminobutyric acid, and ornithine, all of which are formed by the gradual transformations of amino-acids originally present. As can be seen from the table, such acids are indeed found in the fossils recorded there.

The stability of each of the amino-acids and the rate of their disappearance can be ascertained through laboratory studies. Published data are fragmentary but this knowledge is gradually being accumulated. Some years ago (Abelson, 1954) it was shown by heating solutions of the amino-acids at high temperature (225–250° C.) that alanine, glutamic acid, glycine, isoleucine, leucine, proline, and valine are the most stable, more so than threonine, lysine, tyrosine, and phenylalanine. Serine and aspartic acid disappear still more rapidly. Some of these amino-acids give rise to others when they decompose (Vallentyne, 1964; Hare and Mitterer, 1967). Aspartic acid, for instance, breaks down into β-alanine, glutamic acid into γ-aminobutyric acid, and arginine into ornithine. In other instances the acids formed are other naturally occurring acids; thus threonine transforms into glycine, and serine into glycine plus alanine. One can expect the high stabilities of the last two acids to make them conspicuous among the free acids in fossils, and reference to *Table XVIII* shows that this is indeed true (Hare and Mitterer, 1969).

The investigation of the fate of an amino-acid is complicated by the presence of other acids as well as by such external factors as moisture and acidity. A thorough study of the system provided by these acids, dry and with water added, is, however, necessary for a full interpretation of the way they disappear from fossils. The simplest aspect of such a study is the determination of the rate at which each amino-acid breaks down on heating a dilute solution. This has been done for alanine. Several years ago (Hare and Mitterer, 1969) it was shown that the breakdown follows the course of a monomolecular, first-order reaction. Extrapolations from the rate at which it takes place at high temperatures indicate that at 100° C. alanine would have a half-life of about 10,000 years, compared to billions of years at room temperature. Estimates based on similar heating experiments have been made for glutamic acid, phenylalanine, threonine, and serine (Vallentyne, 1964).

Moist and dry amino-acids respond differently to heat. When moist, glycine for instance alters to a dark brown, highly soluble material. Heated at 140° C. or higher after a preliminary drying at 110° C., it darkens and gradually becomes insoluble in water, though for some time it can be hydrolysed to furnish a large but diminishing quantity of glycine.

When mixtures of amino-acids are heated, in solution or as solids, the less stable will disappear while the rest will interact in various ways. Thus when a dilute equimolecular solution of

*Table XIX.*—COMPOSITIONS OF TYPICAL SOLIDS FORMED BY HEATING AMINO-ACID MIXTURES (MOLE PER CENT)*

| AMINO-ACID | 120° C. | 140° C. |
|---|---|---|
| Hypro | 2·3 | 2·0 |
| Asp | 6·7 | 1·8 |
| Thr | 4·6 | 0 |
| Ser | 11·1 | 0 |
| Glu | 12·5 | 5·1 |
| Pro | 7·4 | 7·2 |
| Gly | 17·6 | 13·9 |
| Ala | 8·7 | 16·8 |
| Val | 5·8 | 11·8 |
| Met | 0·8 | 0 |
| Ileu | 2·0 | 7·0 |
| Allo-ileu | 0 | 8·6 |
| Leu | 5·0 | 8·0 |
| Tyr | 1·7 | 0 |
| Phe | 9·2 | 2·9 |
| Orn | 0 | 3·5 |
| Lys | 3·2 | 3·2 |
| Arg | 0 | 0 |
| γ-ABA | 0 | 8·2 |

* Analyses from this laboratory: unpublished data.

all the amino-acids encountered in fossils is heated in a closed vessel at 120–130° C., the less stable acids disintegrate and there is an actual increase in the amounts of glycine and alanine. When the heating is prolonged at these temperatures, and more rapidly at 140° C., a white solid often precipitates. This condensate hydrolyses with strong HCl to yield the more stable amino-acids of the heated mixture (*Table XIX*). Though possessing considerable thermal stability, after several weeks of heating it no longer hydrolyses. The temperature range for its formation is strictly limited, for it has not been obtained at either 110° C. or at 200° C., nor does it always precipitate within these limits. The condensates made thus far have varied in composition from experiment to experiment; none has resembled the proteinaceous residues from fossils.

These solids obtained with dilute amino-acid solutions are in some respects similar to those prepared by heating mixtures of the acids themselves at 170° C. (Fox and Harada, 1960). The products were soluble in salt solutions, hydrolysed with strong acid to yield amino-acids, and had large molecular weights and some of the other properties of simple proteins. Because they may perhaps be related to the proteins of early forms of life, these so-called proteinoids have already been the objects of much study (reviewed in Fox and Nakashima, 1967). There is some evidence (Steinman, 1967) that the sequence of amino-acid residues composing them may not be random but is determined by the reactivities and abundance of the acids from which they are made. Evidently further investigation of condensates is desirable, not only because they could conceivably have been a sub-stratum of primitive life but also to see if they are related in one way or another to the fossil proteins.

## THE PROTEINS OF INVERTEBRATE FOSSILS

Since proteinaceous residues, and not merely their freed amino-acids, are present in exceedingly old vertebrate fossils, it seems reasonable to expect that similar substances will be recovered from fossil shells of the still more ancient, primitive invertebrates. These shells are like those of vertebrate eggs in being proteinaceous membranes embedded in layers of crystalline calcium carbonate, and their compact structure should favour the preservation of the organic component. Much is known about shells through extensive studies of numerous Modern invertebrate species, and many analyses of their proteins have been carried out. More than one protein appears to be associated with their often highly complex structures. Some shells have evolved through palaeontologically demonstrable steps from more primitive forms while others have retained their original character with little apparent change for hundreds of millions of years.

A few early observations based on paper chromatography were made of the proteinaceous residues from fresh and fossil shells. Following the introduction of more nearly quantitative techniques, analyses (for instance, Degens and Spencer, 1966) have been made of the total protein in the shells of a hundred or more living invertebrates, but the presence of somewhat different substances in the same shell means that much remains to be done. Nevertheless, species, generic, and family relationships have been noted between the proteins of the various invertebrates, and it will be important to see the extent to which these relationships persist in the residues isolated from fossils.

Among Modern invertebrates, mollusc shells have been most extensively analysed for their proteinaceous contents. Averages for the proteins of nacreous structures in many gastropod, pelecypod, and cephalopod shells are stated (Hare and Abelson, 1965) in *Table XX*. The gastropod proteins are far richer than the

*Table XX*.—AVERAGE COMPOSITION OF THE NACREOUS PROTEINS OF THREE CLASSES OF MODERN MOLLUSCS (MOLE PER CENT)*

| AMINO-ACID | GASTROPODS | PELECYPODS | CEPHALOPODS |
|---|---|---|---|
| Asp | 10–19 | 8–10 | 8–10 |
| Thr | 1–3 | 1–2 | 0–1 |
| Ser | 8–11 | 6–10 | 9–10 |
| Glu | 5–7 | 2–4 | 5–7 |
| Pro | 2–4 | 0–3 | 1–2 |
| Gly | 18–23 | 29–35 | 29–32 |
| Ala | 16–22 | 23–27 | 23–24 |
| Val | 3–5 | 2–3 | 1–2 |
| ½ Cys | 0–2 | 0–1 | 0–1 |
| Met | 1 | 0–1 | 0–1 |
| Ileu | 2 | 2 | 2 |
| Leu | 4 | 4–6 | 2–3 |
| Tyr | 2–4 | 0–2 | 0–1 |
| Phe | 2–5 | 1–3 | 6 |
| His | 0–1 | 0–1 | 0–1 |
| Lys | 1–3 | 1–2 | 0–1 |
| Arg | 5 | 2–3 | 4 |

* Numerical data estimated from Hare and Abelson (1965).

others in aspartic acid and poorer in both glycine and alanine. Pelecypod and cephalopod proteins are much alike except for larger amounts of glutamic acid and phenylalanine in the latter. The differences that have been observed between nacreous and non-nacreous proteins of gastropods are shown in *Table XXI*; the most striking is in the content of alanine. *Table XXII* demonstrates that even greater differences exist between the various

*Table XXI.*—AVERAGE COMPOSITION OF RECENT GASTROPOD PROTEINS (MOLE PER CENT)*

| AMINO-ACID | NON-NACREOUS | | | NACREOUS |
|---|---|---|---|---|
| | Archaeo-gastropoda | Mesogas-tropoda | Neogas-tropoda | |
| Asp | 14–25 | 9–15 | 10–13 | 11–20 |
| Thr | 4–6 | 4–7 | 4–6 | 2–3 |
| Ser | 6–10 | 4–7 | 4–8 | 8–11 |
| Glu | 6–10 | 9–16 | 8–11 | 5–7 |
| Pro | 8–10 | 5–8 | 6–9 | 2–4 |
| Gly | 14–19 | 10–15 | 10–15 | 18–23 |
| Ala | 6–10 | 7–12 | 8–9 | 16–23 |
| Val | 4–5 | 5–6 | 5–6 | 3–4 |
| ½ Cys | 0–1 | 0–1 | 0–2 | 0–2 |
| Met | 1–6 | 2–7 | 3–9 | 0–2 |
| Ileu | 3–5 | 3–5 | 3–4 | 1–2 |
| Leu | 4–7 | 7–15 | 7–10 | 3–4 |
| Tyr | 2–4 | 1–3 | 2–3 | 2–4 |
| Phe | 2–3 | 3–5 | 3–5 | 2–4 |
| His | 0–1 | 0–1 | 1–2 | 0–1 |
| Lys | 2–4 | 1–3 | 2–3 | 1–3 |
| Arg | 3–5 | 3–6 | 4–6 | 5–6 |

* Numerical data estimated from Hare and Abelson (1965).

*Table XXII.*—AVERAGE COMPOSITION OF PROTEINS OF TWO FAMILIES OF RECENT PELECYPODS (MOLE PER CENT)*

| AMINO-ACID | VENERIDAE | MYTILIDAE |
|---|---|---|
| Asp | 19–28 | 8–10 |
| Thr | 3–5 | 0–1 |
| Ser | 5–7 | 6–10 |
| Glu | 7–9 | 3–5 |
| Pro | 7–9 | 0–3 |
| Gly | 11–15 | 28–35 |
| Ala | 4–6 | 23–28 |
| Val | 3–5 | 2 |
| ½ Cys | 1 | 0–1 |
| Met | 1 | 0–1 |
| Ileu | 3–5 | 2 |
| Leu | 4 | 5–7 |
| Tyr | 5 | 0–3 |
| Phe | 4 | 1–3 |
| His | 1 | 0 |
| Lys | 3–6 | 1 |
| Arg | 3–5 | 2 |

* Numerical data estimated from Hare and Abelson (1965).

*Table XXIII.*—Average Composition of Proteins of Three Recent Veneridae Genera (Mole per cent)*

| Amino-acid | *Mercenaria mercenaria* | *Protothaca grata* | *Tivella argentina* |
|---|---|---|---|
| Asp | 20–21 | 24–25 | 28–30 |
| Thr | 4–5 | 5 | 3–4 |
| Ser | 6–7 | 6 | 6 |
| Glu | 8–9 | 9 | 7–9 |
| Pro | 9 | 7–8 | 9 |
| Gly | 11 | 15 | 13–14 |
| Ala | 6–7 | 5 | 5–6 |
| Val | 3–4 | 5 | 6 |
| ½ Cys | 1–2 | 2 | 1–2 |
| Met | 2 | 2 | 1–2 |
| Ileu | 3 | 4 | 4 |
| Leu | 4 | 4 | 4 |
| Tyr | 5 | 5 | 4–5 |
| Phe | 3 | 3 | 3 |
| His | 1 | 1 | 1 |
| Lys | 5–6 | 4–5 | 4 |
| Arg | 5 | 4–5 | 4 |

* Numerical data estimated from Hare and Abelson (1965).

*Table XXIV.*—Average Composition of Proteins of Three Recent *Mytilus* Species (Mole per cent)*

| Amino-acid | *Mytilus californianus* | *Edulis* | *Veridus* |
|---|---|---|---|
| Asp | 10 | 10 | 9 |
| Thr | 1 | 1 | 1 |
| Ser | 10 | 10 | 8 |
| Glu | 3 | 4 | 3 |
| Pro | 1 | 1 | 0–1 |
| Gly | 29–30 | 29–30 | 34–35 |
| Ala | 26–27 | 26–27 | 26 |
| Val | 2 | 2 | 2 |
| ½ Cys | 1 | 1 | 0–1 |
| Met | 0–1 | 0–1 | 0–1 |
| Ileu | 1 | 1 | 1 |
| Leu | 4–5 | 4–5 | 6 |
| Tyr | 2 | 2 | 0–1 |
| Phe | 2 | 2 | 3 |
| His | 0–1 | 0–1 | 0–1 |
| Lys | 1–2 | 1–2 | 1 |
| Arg | 3 | 3 | 3 |

* Numerical data estimated from Hare and Abelson (1965).

*Table XXV*.—COMPOSITION OF INSOLUBLE PROTEINS IN PLEISTOCENE PELECYPOD SHELLS (MOLE PER CENT)*

| AMINO-ACID | OSTREIDAE | ARCIDAE | CARDIIDAE | | VENERIDAE | | | | |
|---|---|---|---|---|---|---|---|---|---|
| | 16989 | 16990 | 16994 | 16998 | 163693 | 16996 | 16997 | 16999 | 16995 |
| Met-S | 0·1 | 0·1 | 0·4 | 0·2 | — | Trace | 0·1 | Trace | — |
| Asp | 25·0 | 19·6 | 11·3 | 15·1 | 12·6 | 16·1 | 13·9 | 16·4 | 9·8 |
| Thr | 3·3 | 3·7 | 2·9 | 4·1 | 5·5 | 3·5 | 2·8 | 3·9 | 4·4 |
| Ser | 3·7 | 3·4 | 4·0 | 3·2 | 5·3 | 2·7 | 2·7 | 4·2 | 8·7 |
| Glu | 9·4 | 9·8 | 8·0 | 9·5 | 10·6 | 10·4 | 9·1 | 12·0 | 10·5 |
| Pro | 4·3 | 7·0 | 10·9 | 11·8 | 7·1 | 8·4 | 12·3 | 7·8 | 8·8 |
| Gly | 21·3 | 10·8 | 24·0 | 8·1 | 11·3 | 8·0 | 8·0 | 10·5 | 13·9 |
| Ala | 7·6 | 9·2 | 9·3 | 6·6 | 9·7 | 5·2 | 6·8 | 5·8 | 6·7 |
| Val | 4·6 | 6·6 | 3·8 | 5·8 | 5·6 | 7·3 | 6·9 | 6·0 | 6·4 |
| ½ Cys | Trace | 2·0 | Trace | 3·6 | 2·3 | 3·4 | 2·5 | 2·7 | 3·6 |
| Met | 0·8 | 1·2 | 0·6 | 2·0 | 1·1 | 1·3 | 1·3 | 2·1 | 0·7 |
| Ileu | 2·8 | 4·6 | 2·0 | 3·6 | 3·8 | 4·9 | 4·2 | 4·4 | 3·6 |
| Leu | 5·3 | 8·1 | 5·5 | 4·8 | 7·2 | 4·9 | 5·7 | 5·8 | 7·0 |
| Tyr | 1·9 | 1·9 | 0·6 | 4·5 | 3·2 | 6·1 | 6·0 | 3·6 | 1·7 |
| Phe | — | 3·4 | — | 4·0 | 3·9 | 5·8 | 4·3 | 3·4 | 2·9 |
| Orn | 0·5 | 0·6 | 0·8 | 0·6 | 0·4 | 0·9 | 1·2 | 0·6 | 1·0 |
| His | 0·5 | 1·1 | 0·6 | 1·0 | 0·8 | 0·9 | 0·5 | 1·0 | 0·9 |
| Lys | 5·4 | 4·2 | 3·5 | 7·1 | 4·8 | 6·6 | 7·3 | 5·8 | 5·6 |
| Arg | 3·5 | 3·0 | 4·9 | 4·4 | 4·8 | 4·2 | 4·6 | 4·1 | 3·9 |
| Total (μg. per g.) | 91 | 28 | 33 | 128 | 148 | 46 | 43 | 106 | 16 |

* Analyses from this laboratory: Matter, Davidson, and Wyckoff (1971).

Families of a single Class (the pelecypods); the Veneridae and Mytilidae examined (Hare, 1963; Hare and Abelson, 1965) are completely unlike in their contents of aspartic acid, glycine, and alanine. Even between the Genera of a single Family (*Table XXIII*) and the Species of a single Genus (*Table XXIV*) (Hare and Abelson, 1965) there are well-marked dissimilarities in the amounts of one or more of the amino-acids.

Analyses have now been made of the insoluble proteins recovered from a considerable number of fossil shells. Results have been obtained from several Pleistocene pelecypods (*Table XXV*) and a few corals and gastropods (*Table XXVI*). Unlike

*Table XXVI.*—COMPOSITION OF VARIOUS PLEISTOCENE SHELL PROTEINS (MOLE PER CENT)*

| AMINO-ACID | COELENTERATES | | | GASTROPODS | |
|---|---|---|---|---|---|
| | Various Corals I7018 | Diploria I7021 | Siderastrea I7022 | Naticidae I69100 | Busycon I69101 |
| Asp | 13·5 | 11·5 | 12·2 | 13·6 | 10·9 |
| Thr | 6·0 | 6·3 | 5·5 | 4·5 | 4·4 |
| Ser | 5·5 | 6·1 | 6·8 | 5·1 | 5·1 |
| Glu | 9·7 | 9·9 | 9·9 | 12·0 | 11·6 |
| Pro | 5·9 | 4·9 | 5·3 | 6·8 | 5·9 |
| Gly | 11·1 | 10·0 | 12·0 | 12·6 | 11·5 |
| Ala | 8·7 | 10·3 | 8·9 | 8·0 | 8·4 |
| Val | 6·8 | 5·9 | 6·3 | 5·3 | 6·0 |
| ½ Cys | 0·8 | 0·2 | 1·2 | 1·5 | 1·6 |
| Met | 1·3 | 2·3 | 1·1 | 4·0 | 3·2 |
| Ileu | 4·9 | 4·8 | 4·4 | 3·6 | 4·1 |
| Leu | 7·9 | 8·4 | 6·9 | 8·4 | 9·9 |
| Tyr | 3·9 | 4·5 | 5·7 | 2·1 | 2·6 |
| Phe | 5·1 | 4·7 | 4·2 | 3·5 | 4·1 |
| Orn | 0·5 | — | 0·5 | 0·5 | 0·5 |
| His | 0·5 | 1·1 | 1·3 | 0·9 | 1·2 |
| Lys | 3·9 | 4·2 | 4·2 | 4·1 | 4·4 |
| Arg | 3·9 | 4·9 | 3·7 | 3·4 | 4·7 |
| Total (μg. per g.) | 73 | 135 | 16 | 25 | 53 |

* Analyses from this laboratory: Matter, Davidson, and Wyckoff (1971).

many of the better preserved bones and teeth of similar age, they usually have lost at least 90 per cent of their original protein. In some instances the remainder has shown amino-acid ratios close to those in the proteins of related Recent shells; in others this is not the case. Several years ago (Degens and Spencer, 1966) the analyses listed in *Table XXVII* were published for a number

*Table XXVII.*—COMPOSITION OF RESIDUES FROM SEVERAL TERTIARY GASTROPODS (MOLE PER CENT)*

| AMINO-ACID | *Supremus revertens* | *Oxystoma* | *Trochiformis* | *Planor biformis* | *Sulcatus* | *Tenuis* |
|---|---|---|---|---|---|---|
| Asp | 7·29 | 7·28 | 9·17 | 8·57 | 8·35 | 6·08 |
| Thr | 2·06 | 2·14 | 1·93 | 3·19 | 2·38 | 3·11 |
| Ser | 7·96 | 5·57 | 7·73 | 7·48 | 5·84 | 6·49 |
| Glu | 12·27 | 13·28 | 10·14 | 10·97 | 15·44 | 14·60 |
| Pro | 8·29 | 5·57 | 6·28 | 2·39 | 2·50 | 5·95 |
| Gly | 12·27 | 14·57 | 17·87 | 12·96 | 18·36 | 12·16 |
| Ala | 8·95 | 10·71 | 15·45 | 11·17 | 15·44 | 9·05 |
| Val | 5·64 | 4·71 | 3·81 | 5·58 | 4·05 | 5·13 |
| ½ Cys | 1·66 | 2·14 | 1·06 | 2·49 | 3·38 | 4·46 |
| Met | 1·66 | 1·07 | 1·16 | 2·79 | 0·67 | 2·16 |
| Ileu | 2·85 | 3·86 | 1·54 | 3·99 | 1·96 | 0 |
| Leu | 4·64 | 6·00 | 3·24 | 5·58 | 3·55 | 5·13 |
| Tyr | 2·75 | 1·67 | 0·87 | 1·59 | 1·88 | 1·62 |
| Phe | 2·98 | 3·00 | 1·74 | 4·29 | 2·38 | 4·73 |
| His | 2·85 | 2·14 | 2·32 | 1·99 | 3·26 | 4·05 |
| Lys | 5·97 | 6·43 | 3·62 | 4·09 | 3·46 | 5·95 |
| Arg | 9·91 | 9·85 | 12·07 | 10·87 | 7·10 | 9·32 |
| Total (μg. per g.) | 382 | 292 | 249 | 126 | 288 | 94 |

* Analytical data from Degens and Spencer (1966).

of Tertiary gastropod fossils. They agree with the two Pleistocene gastropod proteins of *Table XXVI* in being rich in glutamic acid and glycine, but some of their amino-acid ratios vary considerably from fossil to fossil and they are surprisingly rich in total retained protein. Partly because, as *Table XXI* has shown, there are important differences between the nacreous and prismatic proteins of Modern gastropods, the meaning of these differences is not now apparent.

When, as in *Table XXVIII*, the proteins of progressively older fossils are compared (Akiyama and others, 1971) certain trends in composition are suggested even though the specimens are not closely related. These trends become clearly marked when the proteins in related shells of increasing age are analysed. The compositions of the proteinaceous residues from oyster shells of ages through the Cretaceous (Matter, Davidson, and Wyckoff, 1969) are shown in *Table XXIX*. As these data indicate, the Eocene and older proteins are practically identical, with a composition that is approached in stages by those that are younger. During this approach the high contents of aspartic acid and

*Table XXVIII*.—Composition of Insoluble Residues of Fossil Shells of Various Ages (Mole per cent)*

| Amino-acid | Pleistocene 16994 | Pliocene 16962 | Miocene 16961 | Miocene 16960 | Eocene 16966 | Cretaceous 16941 |
|---|---|---|---|---|---|---|
| Hypro | 6·3 | 0·3 | 0·4 | 0·5 | 0·3 | 0·2 |
| Asp | 10·6 | 8·1 | 11·5 | 12·3 | 8·6 | 7·3 |
| Thr | 3·4 | 4·1 | 6·1 | 5·3 | 5·2 | 5·4 |
| Ser | 4·7 | 5·3 | 8·0 | 5·9 | 10·2 | 12·9 |
| Glu | 8·3 | 9·5 | 9·7 | 10·7 | 13·5 | 11·4 |
| Pro | 10·9 | 5·7 | 5·0 | 4·5 | 6·7 | 4·2 |
| Gly | 24·6 | 10·3 | 14·0 | 15·7 | 16·5 | 22·1 |
| Ala | 9·1 | 9·7 | 11·3 | 13·3 | 7·8 | 6·7 |
| Val | 3·7 | 8·9 | 6·1 | 6·1 | 5·7 | 5·4 |
| ½ Cys | Trace | 0·1 | 1·4 | 1·7 | 1·2 | 1·2 |
| Met | 0·5 | 1·2 | 0·9 | 0·6 | 1·0 | 0·6 |
| Ileu | 2·1 | 7·9 | 3·9 | 3·8 | 4·0 | 3·7 |
| Allo-ileu | — | 1·6 | 0·1 | 0·1 | — | Trace |
| Leu | 4·9 | 13·4 | 8·1 | 8·9 | 6·7 | 7·0 |
| Tyr | 0·9 | 2·4 | 2·2 | 2·4 | 2·5 | 2·6 |
| Phe | 2·2 | 4·2 | 3·5 | 3·3 | 3·6 | 2·7 |
| Orn | 0·8 | 0·9 | 0·4 | 0·5 | 0·4 | 0·3 |
| His | 0·6 | 0·8 | 1·1 | 0·6 | 1·4 | 1·1 |
| Lys | 3·5 | 3·5 | 4·4 | 3·2 | 3·4 | 4·2 |
| Arg | 4·6 | 3·1 | 5·5 | 3·4 | 4·3 | 4·3 |
| β-Ala | — | Trace | 0·4 | 0·5 | 0·5 | 0·2 |
| γ-ABA | — | 0·6 | 0·4 | 0·2 | 0·5 | — |
| Total (μg. per g.) | 32·6 | 12·0 | 25·0 | 13·8 | 5·1 | 16·7 |

* Analyses from this laboratory: Akiyama, Davidson, Matter, and Wyckoff (1971).

glycine in the Recent oyster shells decline with compensatory increases in other amino-acids, notably glutamic acid and leucine.

The same sort of trend is apparent in analyses of a series of scallop shells of ages back through the Jurassic (Akiyama and Wyckoff, 1970; Akiyama, 1971). The insoluble proteinaceous residues older than the Pleistocene are of similar compositions (*Table XXX*), though for some as yet inexplicable reason the Cretaceous specimen departs from the others in its low aspartic and glutamic acid contents and its larger amount of leucine. A similar shift in ratios occurs in the total amino-acid fractions of

*Table XXIX.*—COMPOSITION OF INSOLUBLE RESIDUES FROM FOSSIL OYSTER SHELLS (MOLE PER CENT)*

| AMINO-ACID | PLEISTOCENE 16932 | PLIOCENE 16933 | OLIGOCENE 16935 | EOCENE 16937 | PALAEOCENE 16938 | CRETACEOUS 16941 |
|---|---|---|---|---|---|---|
| Asp | 27·5 | 21·3 | 10·1 | 8·2 | 8·7 | 7·3 |
| Thr | 1·4 | 2·2 | 5·0 | 2·9 | 2·5 | 6·2 |
| Ser | 9·1 | 7·0 | 9·8 | 13·9 | 12·7 | 10·9 |
| Glu | 5·5 | 10·1 | 13·4 | 12·7 | 12·6 | 12·4 |
| Pro | 2·6 | 2·8 | 3·2 | 3·6 | 4·9 | 3·8 |
| Gly | 32·0 | 23·5 | 19·4 | 19·0 | 16·9 | 20·9 |
| Ala | 3·8 | 3·7 | 7·3 | 5·0 | 7·4 | 6·3 |
| Val | 2·8 | 4·2 | 5·1 | 4·9 | 6·0 | 5·7 |
| ½ Cys | 2·2 | 3·0 | 0·3 | 1·4 | 1·2 | 1·2 |
| Met | 0·5 | 0·6 | 0·1 | 1·1 | 0·7 | 0·4 |
| Ileu | 1·7 | 3·0 | 3·3 | 3·8 | 3·3 | 3·0 |
| Leu | 2·8 | 5·6 | 6·4 | 7·4 | 7·3 | 6·7 |
| Tyr | 2·2 | 2·8 | 3·6 | 2·5 | 1·9 | 2·6 |
| Phe | 1·4 | 2·4 | 2·5 | 2·8 | 2·9 | 2·3 |
| Orn | 0·1 | 0·3 | 0·2 | 0·1 | 0·2 | 0·3 |
| His | 0·3 | 0·9 | 1·1 | 1·5 | 1·2 | 1·1 |
| Lys | 2·0 | 3·4 | 4·8 | 4·4 | 3·9 | 4·6 |
| Arg | 2·3 | 3·5 | 4·4 | 4·9 | 5·7 | 4·3 |
| Total (μg. per g.) | 1231·0 | 67·7 | 20·1 | 9·8 | 19·3 | 16·7 |

* Analyses from this laboratory: Matter, Davidson, and Wyckoff (1969).

*Mercenaria* shells of ages receding into the Miocene (Hare and Mitterer, 1967) but with them (*Table XXXI*) there are precipitate declines in aspartic acid and serine but not in glycine, with balancing increases in glutamic acid, alanine, and valine but not in leucine. It is to be noted that though each of these three groups of shells shows an approach with age to a common product, the composition of these end-products is different in each case. The amount is small but surprisingly constant, there being nearly as much in a Cretaceous or Jurassic shell as in many that are no older than the Pleistocene. Evidently most of the protein of a fresh shell disappears quickly but there remains either a very resistant component or a product of the decomposition of the unstable protein which, when formed, is very resistant to change.

In another investigation (Bricteux-Grégoire, Florkin, and Grégoire, 1968) the proteins from the prismatic region of an Eocene and a Cretaceous *Pinna* shell were compared with both the prismatic and the nacreous proteins of Modern relatives. The two nacreous proteins are very different in composition from the Modern prismatic proteins (*Table XXXII*), which in turn are

*Table XXX.*—Insoluble Proteins in Recent and Fossil Scallops (Mole per cent)*

| Amino-acid | Recent | | Pleistocene† | Pliocene‡ | Miocene§ | Oligocene I7032 | Cretaceous I7033 | Jurassic I709 |
|---|---|---|---|---|---|---|---|---|
| Asp | 30·8 | 28·6 | 13·0 | 11·7 | 11·1 | 10·3 | 1·3 | 7·4 |
| Thr | 1·7 | 2·1 | 3·7 | 4·7 | 6·2 | 4·8 | 7·0 | 5·6 |
| Ser | 21·3 | 20·7 | 9·1 | 10·6 | 8·7 | 9·2 | 8·3 | 9·7 |
| Glu | 5·3 | 6·2 | 11·1 | 10·0 | 10·1 | 10·8 | 3·6 | 11·2 |
| Pro | 1·1 | 2·5 | 4·2 | 5·8 | 5·6 | 4·3 | 6·0 | 5·3 |
| Gly | 22·7 | 22·2 | 18·2 | 17·6 | 18·3 | 16·8 | 16·8 | 17·5 |
| Ala | 6·3 | 6·4 | 11·0 | 10·1 | 11·2 | 12·2 | 15·7 | 11·7 |
| Val | 1·9 | 1·6 | 6·3 | 6·0 | 5·7 | 5·9 | 9·8 | 6·9 |
| ½ Cys | 1·0 | 1·0 | 0·1 | Trace | 0 | 0·4 | 0 | 0·6 |
| Met | 0·4 | 1·9 | 0·6 | 1·0 | 0·5 | 1·6 | 0·7 | 2·9 |
| Ileu | 1·1 | 1·5 | 3·8 | 4·5 | 3·8 | 3·3 | 5·8 | 3·7 |
| Allo-ileu | 0 | 0 | 0·6 | 0·4 | 0·1 | 0·2 | 1·1 | Trace |
| Leu | 2·2 | 2·7 | 8·6 | 8·7 | 8·6 | 8·7 | 12·9 | 8·0 |
| Tyr | 0 | 0 | 2·6 | 1·8 | 2·4 | 2·7 | 1·4 | 2·7 |
| Phe | 0·6 | 0·9 | 3·1 | 2·9 | 3·7 | 3·3 | 4·5 | 3·4 |
| Hylys | 0·1 | 0·1 | 0 | 0 | 0 | 0 | 0 | 0 |
| Orn | 0 | 0 | 0·5 | 0·4 | 0·1 | Trace | 0·7 | 0 |
| His | 0·2 | 0·1 | 0 | 0 | 0 | 0 | 0 | 0 |
| Lys | 2·8 | 1·3 | 3·2 | 3·3 | 3·2 | 2·5 | 2·6 | 2·5 |
| Arg | 0·4 | 0·2 | 0 | 0 | 0 | 0 | 0 | 0 |
| β-Ala | 0 | 0 | 0·4 | 0·6 | 0·4 | 1·9 | 1·1 | 0·4 |
| γ-ABA | 0 | 0 | Trace | 0·1 | 0·4 | 1·3 | 0·9 | 0·7 |
| Total (μg. per g.) | 1399 | 2887 | 19·6 | 9·7 | 25·0 | 12·1 | 7·7 | 4·3 |

* Analyses from this laboratory: Akiyama and Wyckoff (1970), Akiyama (1971).
† Averaged from Samples I7027, I7046, I7048, I7049, I7050.
‡ Averaged from Samples I7029, I7030, I7031, I6959.
§ Averaged from Samples I67114, I6960, I6961.

8

unlike the fossil (prismatic) proteins; in the latter, serine and glutamic acid have greatly increased in importance at the same time that the amount of glycine has lessened. There is more alanine in the Eocene *Pinna* but otherwise the two fossil proteins are similar and, except for a significantly greater amount of

*Table XXXI.*—COMPOSITION OF TOTAL SOLUBLE FRACTIONS FROM *Mercenaria* SHELLS (MOLE PER CENT)*

| AMINO-ACID | MODERN | 1000 YEARS OLD | UPPER PLEISTO-CENE | PLEISTO-CENE | PLIOCENE | MIOCENE | |
|---|---|---|---|---|---|---|---|
| | | | | | | Upper | Middle |
| Asp | 29·54 | 26·41 | 24·02 | 16·58 | 13·78 | 11·38 | 4·17 |
| Thr | 3·56 | 2·92 | 1·94 | 1·01 | 0·46 | 0·07 | 0·24 |
| Ser | 12·12 | 5·05 | 2·49 | 0·47 | 2·03 | 0·54 | 1·47 |
| Glu | 5·84 | 9·56 | 9·98 | 10·53 | 12·76 | 11·72 | 14·70 |
| Pro | 6·85 | 13·27 | 14·36 | 13·30 | 14·25 | 16·21 | 16·67 |
| Gly | 15·33 | 12·47 | 8·91 | 10·10 | 14·71 | 10·90 | 13·23 |
| Ala | 4·69 | 6·81 | 9·98 | 14·05 | 17·21 | 19·07 | 19·36 |
| Val | 2·09 | 3·46 | 3·88 | 6·28 | 6·10 | 7·42 | 7·35 |
| Met | 0·17 | 0·45 | 0·69 | 0·78 | Trace | 0·41 | 0·24 |
| Ileu | 1·87 | 2·03 | 1·66 | 2·07 | 2·03 | 2·11 | 1·96 |
| Allo-ileu | 0 | 0·31 | 0·88 | 1·99 | 2·40 | 2·79 | 2·70 |
| Leu | 3·03 | 2·88 | 3·14 | 4·49 | 3·42 | 4·09 | 4·41 |
| Tyr | 4·17 | 2·50 | 3·88 | 3·47 | 1·39 | 1·77 | 1·22 |
| Phe | 2·64 | 2·61 | 3·23 | 3·51 | 2·50 | 3·13 | 2·94 |
| Orn | 0 | 0·96 | 1·85 | 2·81 | 2·22 | 2·11 | 1·47 |
| His | 0·42 | 0 | 0 | — | — | — | — |
| Lys | 4·80 | 4·58 | 5·17 | 5·97 | 2·78 | 3·27 | 3·68 |
| Arg | 2·88 | 3·69 | 3·09 | 1·33 | — | — | — |
| γ-ABA | 0 | 0 | 0·55 | 0·23 | 0·74 | 1·57 | 2·45 |
| α-ABA | — | 0 | 0·28 | 1·01 | 1·20 | 1·43 | 1·71 |

* Analytical data calculated from Hare and Mitterer (1967).

aspartic acid, of the same general type as those from the older oysters (*Table XXIX*).

Very recently an extensive series of analyses has been carried through on fossil Nautiloid and Ammonoid shells. A preliminary account has been published (Grégoire and Voss-Foucart, 1970) but through the courtesy of Dr. Grégoire detailed data (to appear in Voss-Foucart and Grégoire, 1971) are reproduced in *Table XXXIII*. The nacreous protein of Modern *Nautilus pompilius* is especially rich in glycine and alanine and in this respect most closely resembles the *Mytilus* proteins of *Table XXIV*. Like the ancient proteins discussed above, the Eocene and older *Nautilus* proteins have practically the same composition which differs from that of the Modern protein and lies more or less midway between those of the ancient oyster and scallop insolubles. These results are especially important in showing that shells as old as the

*Table XXXII.*—COMPOSITION OF MODERN AND FOSSIL *Pinna* PROTEINS (MOLE PER CENT)*

| AMINO-ACID | MODERN | | | | FOSSIL PRISM | |
|---|---|---|---|---|---|---|
| | *Atrina nigra* | | *Pinna nobilis* | | *Pinna affinis* | *Inoceramus* |
| | Nacre | Prism | Nacre | Prism | Eocene | Cretaceous |
| Asp | 12·6 | 10·4 | 13·5 | 23·3 | 11·1 | 10·0 |
| Thr | 1·4 | 1·1 | 1·8 | 1·5 | 4·7 | 4·4 |
| Ser | 9·5 | 4·0 | 9·0 | 4·6 | 16·5 | 16·4 |
| Glu | 4·1 | 1·8 | 4·1 | 1·9 | 13·6 | 15·5 |
| Pro | 2·3 | 2·2 | 2·0 | 2·2 | Trace | Trace |
| Gly | 19·2 | 46·2 | 21·0 | 36·9 | 22·5 | 21·3 |
| Ala | 30·8 | 4·7 | 28·9 | 5·7 | 14·1 | 8·1 |
| Val | 2·5 | 6·3 | 2·4 | 5·6 | 5·0 | 4·9 |
| Ileu | 2·6 | 2·0 | 2·6 | 3·4 | 3·8 | 3·4 |
| Leu | 4·2 | 7·4 | 3·8 | 4·5 | 5·9 | 5·7 |
| Tyr | 1·7 | 9·3 | 1·9 | 5·3 | Trace | Trace |
| Phe | 2·9 | 2·3 | 2·4 | 2·4 | 2·9 | Trace |
| His | Trace | 0·4 | Trace | 0·5 | Trace | 3·2 |
| Lys | 1·8 | 0·8 | 2·3 | 0·6 | Trace | 4·8 |
| Arg | 4·1 | 1·1 | 4·3 | 1·7 | Trace | 2·4 |
| Total (µg. per g.) | — | 1930 | — | 2020 | 96 | 20 |

* Analytical data from Bricteux-Grégoire, Florkin, and Grégoire (1968).

Devonian have retained enough protein for complete analysis and that amino-acids are present in the same relative amounts as in younger shells. The Cretaceous and Jurassic proteins of the related Ammonoids are alike in most respects and substantially the same as those from the older *Nautilus* shells. It is to be noted that the Carboniferous Nautiloid and many of the Cretaceous and some of the Jurassic specimens have not transformed from aragonite to the more stable calcite.

The foregoing analyses of the proteinaceous residues in fossil shells emphasize the generalization, already suggested, that, though there is a wide diversity in the composition of the proteins in fresh shells, the residues from related species of fossils

*Table XXXIII.*—Composition of Insoluble Residues from Nautiloids and Ammonoids (Mole per cent)*

| Amino-acid | Nautiloids | | | | | | | | | | | | |
|---|---|---|---|---|---|---|---|---|---|---|---|---|---|
| | Modern | Miocene | Eocene | | Cretaceous | | | Jurassic | | | Permian | Carboni-ferous | Devonian |
| | *Nautilus pompilius* | *Eutrepho-ceras* | — — | — — | *Cymato-ceras* | *Eutrepho-ceras* | — — | — — | — — | — — | *Domato-ceras* | — — | — — |
| Asp | 7·1 | 7·9 | 9·1 | 8·0 | 7·6 | 9·4 | 10·4 | 9·2 | 8·6 | 9·1 | 8·8 | 9·0 | 8·7 |
| Thr | 1·2 | 2·3 | 4·4 | 4·5 | 4·8 | 5·1 | — | 4·0 | 4·9 | 5·4 | — | 4·6 | 4·5 |
| Ser | 9·6 | 7·6 | 9·1 | 11·4 | 12·6 | 15·8 | 15·5 | 11·4 | 10·1 | 13·0 | 15·1 | 14·4 | 19·4 |
| Glu | 4·4 | 7·9 | 11·1 | 10·8 | 18·4 | 12·4 | 15·2 | 16·7 | 11·6 | 12·1 | 14·5 | 11·3 | 13·3 |
| Pro | 0·7 | 4·0 | 5·0 | — | 7·4 | — | 3·8 | 4·3 | 4·1 | 4·8 | — | 3·3 | — |
| Gly | 31·6 | 24·8 | 18·5 | 20·0 | 15·9 | 18·1 | 17·7 | 18·4 | 20·4 | 18·6 | 21·8 | 19·0 | 17·2 |
| Ala | 26·6 | 8·9 | 9·8 | 8·2 | 7·2 | 8·8 | 9·3 | 8·3 | 7·8 | 7·4 | 6·6 | 7·3 | 7·6 |
| Val | 1·5 | 4·6 | 5·4 | 5·5 | 4·7 | 5·3 | 4·7 | 4·4 | 4·8 | 4·5 | 3·8 | 4·3 | 3·2 |
| Cys | Trace | Trace | Trace | Trace | 0 | Trace | 0 | Trace | 1·5 | Trace | 0 | Trace | 0 |
| Met | 0·5 | Trace | 0 | Trace | Trace | Trace | 0·3 | 1·3 | 0·6 | Trace | 0 | 0·6 | 0 |
| Ileu | 1·4 | 2·6 | 3·4 | 4·0 | 3·4 | 2·7 | 2·9 | 2·4 | 3·4 | 3·2 | 3·6 | 2·9 | 2·0 |
| Leu | 2·1 | 4·8 | 6·0 | 7·8 | 6·3 | 4·8 | 4·9 | 4·7 | 6·5 | 6·0 | 6·6 | 5·5 | 3·1 |
| Tyr | 1·8 | 2·8 | 1·7 | 2·1 | — | 2·3 | 1·9 | 1·1 | 3·1 | 2·5 | 2·8 | 2·1 | 1·3 |
| Phe | 5·1 | 12·1 | 6·1 | 7·9 | — | 2·4 | 2·2 | 3·5 | 3·3 | 3·4 | 3·0 | 3·1 | 1·8 |
| His | 0·5 | 1·6 | 1·7 | 1·2 | 2·2 | 2·4 | 2·2 | 2·8 | 1·3 | 1·6 | 2·7 | 3·0 | 5·3 |
| Lys | 0·6 | 5·5 | 5·4 | 5·3 | 4·7 | 5·9 | 6·0 | 5·6 | 3·5 | 4·1 | 6·9 | 4·8 | 7·8 |
| Arg | 5·3 | 2·5 | 3·4 | 3·4 | 4·9 | 3·5 | 2·8 | 1·9 | 4·5 | 4·2 | 3·8 | 5·0 | 4·6 |
| Mineral† | Ara | Ara | Ara | Ara | Ara | Ara | Cal | Ara | Cal | Ara | Cal | Ara | Cal |

in either composition or amount. In this case the persisting insoluble protein of the Pleistocene shell is like the unheated material except for its loss of alanine. In another group of experiments, thus far only incompletely reported, portions of fresh *Nautilus* shells were heated for several hours *in vacuo* or in air at temperatures up to 900° C. After heating, as before, they were biuret-positive and the decalcified residues still possessed under the electron microscope (Grégoire, 1968) some of the pebbly structure characteristic of the fresh protein membranes. As might be expected, the damage was less severe after the *in vacuo* heating. Amino-acids are not the only products of protein decomposition which are biuret-positive and quantitative amino-acid analyses are evidently required. Approximate values obtained by pyrolizing the nacreous portion of fresh *Nautilus* shells are reported (Grégoire and Voss-Foucart, 1970) in the last column of *Table XXXIII*. They closely resemble the results for the fossil proteins of this table. It would appear from these two sets of experiments that heating fresh shells leads to a proteinaceous product similar in composition to that found in old fossil shells of related species. It is also apparent that though much of the fresh protein of a shell disappears more rapidly than does the collagen of bones, the residue is more stable than collagen unless the latter has remained unusually dry.

The rapid decay of most of the proteins of shells makes it especially necessary to investigate the free amino-acids and peptides that are the products of this decay. Few complete data of this sort exist, but they have been accumulated for a group of scallop (*Pecten*) shells of ages back through the Jurassic. Analyses (Akiyama, 1971) for their free acids and soluble peptides are given in *Tables XXXV* and *XXXVI*. More than 70 per cent of the protein in a Recent shell (*Table XXX*) is made up of glycine, aspartic acid, and serine. In the free-acid fraction from the fossils the relatively unstable serine is absent, its place being taken by the stable alanine. Thus, as with fossil bones, the free-acid fraction is dominated, not so much by those acids that are plentiful in the initial protein, as by the more stable among them. The soluble peptides in these scallop fossils are of an exceptionally simple composition. They are more than half aspartic plus glutamic acids in specimens of ages from the Pleistocene back through the Oligocene; over this interval of time the relative amount of aspartic acid diminishes and of glutamic acid

*Table XXXV*.—Composition of the Free Amino-acids from Fossil Scallops (Mole per cent)*

| Amino-acid | Pleistocene† | Pliocene‡ | Miocene§ | Oligocene I7032 | Cretaceous I7033 | Jurassic I709 |
|---|---|---|---|---|---|---|
| Asp | 36·0 | 33·3 | 29·9 | 12·6 | 35·8 | 12·9 |
| Thr | 0·3 | 0·4 | — | — | 0·5 | — |
| Ser | 1·3 | 0·3 | 0·4 | 0·2 | 1·2 | 2·4 |
| Glu | 4·2 | 4·3 | 4·5 | 6·0 | 8·0 | 7·6 |
| Pro | 2·7 | 3·3 | 2·0 | 2·4 | 1·3 | 2·2 |
| Gly | 17·0 | 15·6 | 12·7 | 10·0 | 11·6 | 22·0 |
| Ala | 23·8 | 27·9 | 38·5 | 56·1 | 18·3 | 31·0 |
| Val | 3·2 | 3·5 | 2·2 | 3·7 | 1·3 | 4·8 |
| CySH | 0·1 | — | 0·1 | 0·2 | Trace | — |
| Met | 1·1 | 0·9 | 1·0 | 1·1 | 0·5 | 2·1 |
| Ileu | 0·8 | 0·6 | 0·3 | 0·4 | 0·3 | 0·6 |
| Allo-ileu | 0·9 | 0·8 | 0·6 | 1·0 | 0·6 | 0·8 |
| Leu | 4·5 | 5·1 | 3·2 | 1·8 | 1·0 | 2·1 |
| Tyr | 0·7 | 0·2 | 0·1 | — | — | 0·3 |
| Phe | 0·8 | 0·7 | 0·5 | 0·4 | 0·3 | 1·0 |
| Orn | 0·6 | 0·4 | 0·3 | — | 2·6 | 2·0 |
| Lys | 0·9 | 1·0 | 0·8 | 0·4 | 1·5 | 1·9 |
| β-Ala | 0·7 | 1·2 | 2·0 | 2·5 | 4·9 | 3·7 |
| γ-ABA | 0·4 | 0·6 | 0·9 | 1·3 | 10·5 | 2·5 |
| Total (μg. per g.) | 224·5 | 144·7 | 149·6 | 35·7 | 10·6 | 7·3 |

* Analyses from this laboratory: Akiyama (1971).
† Averaged from Samples I7027, I7046, I7048, I7049, I7050.
‡ Averaged from Samples I7029, I7030, I7031, I6959.
§ Averaged from Samples I67114, I6960, I6961.

steadily increases with age. In still older shells alanine, and especially glycine, are of greater importance than before. One of the most interesting tasks for the future will be the identification of these apparently simple peptides.

In considering the insoluble residues in fossil shells it was pointed out that in specimens older than Pleistocene the amount was practically independent of age. A different picture emerges for the scallop shells if the total amino-acid content (free acids, plus peptides, plus protein) is related to fossil age (*Table XXXVII*). At least with this group of specimens there is an unbroken decrease with age. In all these shells the free acids and soluble peptides are in larger amounts than the insoluble residues. The quantities vary widely from specimen to specimen but in the

*Table XXXVI.*—Composition of the Soluble Peptides from Fossil Scallops (Mole per cent)*

| Amino-acid | Pleistocene† | Pliocene‡ | Miocene§ | Oligocene I7032 | Cretaceous I7033 | Jurassic I709 |
|---|---|---|---|---|---|---|
| Asp | 61·8 | 44·7 | 27·8 | 16·7 | 20·5 | 10·3 |
| Thr | 0·8 | 0·5 | 0·8 | — | 4·5 | 5·2 |
| Ser | 2·2 | 5·2 | 5·0 | 5·7 | 9·0 | 5·1 |
| Glu | 13·7 | 18·7 | 33·5 | 36·9 | 20·5 | 24·4 |
| Pro | 1·2 | 1·2 | 2·2 | 1·4 | 4·1 | 4·9 |
| Gly | 5·9 | 11·6 | 9·6 | 15·6 | 18·3 | 22·9 |
| Ala | 5·4 | 8·0 | 6·4 | — | 6·6 | 13·7 |
| Val | 2·4 | 1·6 | 2·6 | 3·1 | 2·8 | 0·2 |
| CySH | 0·3 | — | 0·1 | 0·9 | — | — |
| Met | 0·4 | 1·3 | 1·1 | 1·7 | 1·3 | 1·5 |
| Ileu | 0·7 | 0·2 | 1·4 | 2·3 | 1·9 | 1·4 |
| Allo-ileu | 0·3 | — | 0·6 | 0·6 | 0·2 | 0·2 |
| Leu | 2·2 | 1·8 | 3·6 | 4·1 | 5·9 | 4·7 |
| Tyr | 0·4 | 0·1 | 0·4 | — | — | 0·6 |
| Phe | 0·4 | 0·4 | 1·4 | 0·8 | 0·9 | 1·6 |
| Orn | 0·4 | 0·2 | 0·4 | — | — | — |
| Lys | 0·7 | 1·0 | 1·0 | 1·3 | 0·7 | 0·8 |
| β-Ala | 0·3 | 2·6 | 0·7 | — | 1·3 | 0·3 |
| γ-ABA | 0·7 | 1·0 | 1·4 | 8·8 | 1·4 | 2·3 |
| Total (μg. per g.) | 238·3 | 114·4 | 29·0 | 6·9 | 4·1 | 5·2 |

* Analyses from this laboratory: Akiyama, 1971.
† Averaged from Samples I7027, I7046, I7048, I7049, I7050.
‡ Averaged from Samples I7029, I7030, I7031, I6959.
§ Averaged from Samples I67114, I6960, I6961.

*Table XXXVII.*—Relation between Amino-acid Fraction and Geologic Age of Fossil Scallop Shells (μg. per g.)*

| | Insolubles | Soluble Peptides | Free Amino-acids | Total Amino-acids | Per Cent Surviving |
|---|---|---|---|---|---|
| Pleistocene | 29·9 | 238·1 | 197·3 | 465·3 | 21·3 |
| Pliocene | 9·7 | 180·3 | 146·2 | 270·2 | 12·4 |
| Miocene | 24·9 | 28·8 | 150·7 | 204·5 | 9·4 |
| Oligocene | 11·4 | 6·9 | 35·7 | 54·0 | 2·5 |
| Cretaceous | 7·6 | 4·1 | 10·6 | 22·3 | 1·0 |
| Jurassic | 4·4 | 5·2 | 7·3 | 16·9 | 0·8 |

* Data from this laboratory: Akiyama and Wyckoff (1970), Akiyama (1971).

Pleistocene shell V7046, for example, there are six times the free acids, and seven times as much peptides as insoluble protein.

In these scallop shells the compositions of the soluble fractions are very different from that of the insoluble, proteinaceous residue of the same age (*Table XXX*). This is not true for the inner, nacreous, layer of the Pleistocene *Mercenaria* shell of *Table XXXIV*. Except for alanine and arginine its total amino-acid and insoluble fractions are almost identical in composition.

In discussing the soluble products of protein decomposition as found in fossil bones (p. 90) it was pointed out that two factors are at work in eliminating them. One is a leaching when water gains access to a fossil. The large amount of free amino-acids in the Miocene specimens of *Table XXXVII* is clear evidence that under suitable conditions a fossil may be preserved for millions of years from such leaching. The other factor effective in eliminating amino-acids is their slow breakdown into other substances and their transformation into other amino-acids. Direct evidence for this is the appearance (*Tables XXX*, *XXXI*, *XXXIV*, and *XXXV*) of allo-isoleucine, β-alanine, γ-aminobutyric acid, and ornithine in soluble fractions from invertebrate shells as well as from bones and teeth (p. 91).

Attention was drawn to them by the analyses of total soluble fractions from the group of *Mercenaria* shells recorded in *Table XXXI* (Hare and Mitterer, 1967). As the table indicates, they are to be found in all soluble fractions and are most prominent in the older specimens. At the time these analyses were carried out, preliminary studies were being made of the stabilities of dilute solutions of several amino-acids and of their isomeric changes as a function of time. Only laevo (L) rotatory acids occur in proteins and they are liberated when a protein disintegrates. Small amounts of the dextro (D) acids are, however, found among the free acids from fossils, undoubtedly having arisen through a rearrangement of some of the freed L-molecules. This isomeric transformation is readily followed in the laboratory by measuring the optical rotation of a heated solution of an amino-acid. Such experiments show that D-isomers tend to revert, at about the same rate as they are formed, to L-acids, and that therefore, when an L-acid alters, there is a gradual approach to an equilibrium in which the two forms would be present in equal amounts. These rates have not yet been determined for most amino-acids but, when they are known, measurements of the ratios of L- to D-forms will further add to our knowledge of the previous history of a fossil.

Unfortunately, the chemical behaviours of D- and L-isomers are so nearly the same that they are not separated by the usual chromatographic procedures. The presence of D-forms was first demonstrated in the fossil *Mercenaria* shells by digestion of their amino-acids with enzymes that attack only L-molecules (Hare and Abelson, 1968). To determine the D-acids in this way the sample was divided into two portions, in one of which the total amino-acid content was measured. An enzyme added to the other half digested the L-acids and was afterwards removed by adding a protein precipitant such as trichloro-acetic acid; the resultant solution of D-acids could then be analysed in the usual fashion. Recently (Kvenvolden, Peterson, and Brown, 1970) it has been shown in studies of amino-acids extracted from ancient sediments that a chromatographic recognition of D- in the presence of L-isomers can be effected by first making derivatives involving the racemic (DL)-forms. Further work undoubtedly will show whether or not this procedure, though more complicated in principle, is preferable to enzymatic digestion when applied to fossils.

The finding of allo-isoleucine among the free amino-acids of fossils introduces an isomerism in which the two forms are more easily distinguished. Isoleucine, with the formula

$$\begin{array}{ccc} & H & H \\ & | & | \\ NH_2 & C^{(\alpha)} - & C^{(\beta)} - C_2H_5 \\ & | & | \\ & COOH & CH_3 \end{array}$$

has two asymmetric carbon atoms, each permitting an isomeric change. Allo-isoleucine, involving isomerism on the (β) carbon atom, is sufficiently different from isoleucine so that the two chromatographic peaks are clearly separated. Its gradual development in fossils at the expense of isoleucine is therefore readily followed. Reference to the tables shows that little if any of the allo-form is present in insoluble protein residues of even very old fossils, but it is rather consistently found in free-acid and peptide fractions. From this fact it would appear that the molecular rearrangement producing it occurs only after isoleucine has split off from its parent protein molecule.

The transformation of isoleucine into allo-isoleucine is, like other isomerisms, a reversible process. According to laboratory measurements (Hare and Abelson, 1968) the reversion of allo-isoleucine to isoleucine proceeds at a somewhat slower rate, with the result that at equilibrium the ratio of allo-isoleucine to leucine is slightly greater than unity. The *Mercenaria* data of *Table XXXI* indicate that this equilibrium prevails in the Pliocene and older specimens; among the scallop fossils of *Table XXXIV* it has already been reached in the Pleistocene shell. In a very recent study (Kvenvolden and others, 1970) isoleucine–allo-isoleucine ratios have been combined with a knowledge of the extrapolated rates of these isomeric transformations to date marine sediments as old as a million years.

Individual bones and shells large enough for analysis exist for periods back into the Ordovician though it is not yet known with certainty if residues of proteinaceous origin can be recovered from fossils older than the Devonian. Farther back in time, however, fossils become so small that ultimately samples for analysis will consist of more than one individual. Plants formed so large a segment of early terrestrial life that the possibility of their contaminating the smaller animal fossils and their matrices becomes more probable. A few preliminary observations on certain Precambrian algae have suggested the continuing presence of protein (unpublished data) but its source has not yet been established. Most ancient plant fossils, however, show evidence of such profound chemical changes that the survival of their proteins and amino-acids is improbable. Decomposition of masses of plant materials usually leads to the formation of the highly insoluble, ill-defined, carbon-rich substance known as kerogen. This constitutes much of the organic matter in ancient fossiliferous rocks. Amino-acids are not to be extracted from it but they have been reported (Heijkenskjöld and Möllerberg, 1958) from hydrolysates of peats and coal. Marine sediments are currently being investigated for their amino-acids (e.g., Erdman, Marlett, and Hanson, 1956; Degens, Emery, and Reuter, 1963; Clarke, 1967; Kvenvolden and others, 1970), but little is yet known about the far smaller quantities that may be in sedimentary rocks and especially those limestones made up of calcareous microfossils. Initially they must have incorporated much organic matter of both plant and animal origin and, now that sufficiently sensitive methods of analysis are at hand, their detailed study may become one of the more promising aspects of the work outlined here.

## *CHAPTER V*

# THE CARBOHYDRATES AND LIPIDS IN FOSSILS

BESIDES proteins and the immediate products of their decomposition, other organic components of living matter which one may reasonably expect to find preserved in individual fossils are carbohydrates and fats. The chief sources of our fossil fuels, they occur as deposits which are grossly altered in both structure and composition from the masses of living matter that were their origin. Decomposition has proceeded so far that relatively little pertinent to the present study can be deduced from the extensive investigations that have been made of coal and petroleum. It has indeed been possible to trace the steps, from peat through the various soft coals to anthracite, by which great accumulations of dead plants have gradually been carbonized. The development of petroleum is less clear because as a fluid it furnishes no morphological evidence of its fossil origins, and it is usually not found where formed. Nevertheless, many crude oils contain compounds that, as decomposition products of chlorophyll and plant pigments, testify to their biological origin.

### POLYSACCHARIDES

The polysaccharides that are found in individual fossils are usually chitin or cellulose. Cellulose is the universal structural unit of plants but only rarely does it occur in animal remains. Chitin is more widespread among animals; together with associated protein and some calcification, it forms the external support and framework of insects and of such invertebrates as lobsters and crabs. It is also present in fungi and perhaps in small amounts in a few other primitive plants.

In composition, chitin is a polymer of glucosamine which has the formula $C_6H_{11}O_5NH_2$. When hydrolysed with strong acid it yields the amine itself. This amine likewise forms part of mucoidal proteins widespread in the animal kingdom. Thus, though it can under suitable conditions be a measure of the chitin in a fossil, it does not necessarily prove the existence of chitin. Glucosamine gives a characteristic peak on both liquid–column and gas chromatograms and can accordingly be determined without special effort in the course of an amino-acid analysis. The older literature contains numerous references to chitin in fossils, and notably in the carapaces of trilobites. The tests that were applied were not, however, of a nature to have convincingly established

```
        CH2OH                          CH2OH
          |                              |
          C——O                           C——O
H       /H     \     H        H        /H     \     H
C        OH      \   C        C         OH      \   C
| \       |    H  /  |        | \        |    H  /  |
OH  \C————C/        OH        OH  \C————C/        OH
     H    |                        H    |
         NH2                           OH

     Glucosamine                    Glucose
```

its presence. One of the few exceptions to this is the proof of its persistence in insect wings of Eocene age (Abderhalden and Heyns, 1933). This situation should change with the wider use of modern techniques.

As the chief organic component of fossil plants, a detailed consideration of cellulose lies outside the scope of this monograph (for a recent review *see* Vallentyne, 1963). The student of animal fossils may, however, encounter it in coprolites and fossil-bearing rocks and since, like chitin, it is amenable to chromatographic identification and analysis, he must be prepared to deal with it. Native cellulose is generally described as a polymer of glucose, $C_6H_{12}O_6$, in which the number of these units per elementary particle is very high, of the order of a thousand. Other glucose polymers having smaller particles, and the different characteristics that depend on this reduced size, occur naturally or can be obtained by the chemical treatment of cellulose. Such shorter polymers are, for instance, the $\beta$- and $\gamma$-celluloses. Glucose polymers of still lower degrees of polymerization are stored as food by both plants and animals. For plants it is starch, for animals glycogen. All these polymers can be split into simple molecules of sugar by hydrolysis. Though cellulose is essentially a glucose polymer, other sugars are frequently obtained on hydrolysis. This is true, for example, with wood; as much as a third of its hydrolysate is the pentose sugar xylose, $C_5H_{10}O_5$. Similar, mixed polysaccharides are present in both simple and highly organized plants. As these substances progressively degrade to yield particles of lower degrees of polymerization, the fragments become, like the simple sugars, soluble in water. It is possible, at least in principle, to control this degradation, to identify the resulting sugars, and in this way to determine their sequence in the fragments and ultimately in the original cellulosic polysaccharide. These essential structures of plants could thus be dissected in the same way as the proteins are analysed in terms of their amino-acids and peptides. Such a complete polysaccharide analysis, however, is technically in many respects even more difficult than the analysis of proteins, and knowledge remains very limited.

The possibilities for more knowledge are nevertheless so great for the student of fossil plants that preliminary essays have been made. Celluloses have been extracted, for instance, from a number of fossil woods dating back to the Cretaceous, their degrees of polymerization have been measured, and, after hydrolysis, the constituent simple sugars identified (*see* review in Vallentyne, 1963). Several were recovered from Miocene wood studied in this fashion. It is an important conclusion of this earlier work that cellulosic chains easily split and that in consequence all the fossil specimens may be expected to have greatly reduced degrees of polymerization.

Polysaccharides with elementary particles smaller than those of cellulose have rarely been sought in fossils and their tendency to break down into soluble sugars makes it improbable that they would be found in any but fairly recent specimens. The reported recovery of immunologically active glucoproteins in mummies (p. 71) nevertheless suggests that the search for them in Pleistocene material might well be richly rewarding.

## Lipids

For the most part the fatty compounds synthesized by all animals and plants do not have the chemical stability to endure for geologically significant periods of time. Some of the characteristic products of their decomposition are, nevertheless, so stable that they provide the most persisting indices of earlier life.

The most conspicuous group of natural fats, the glycerides, are esters of the polyalcohol glycerin. They have the chemical formula:

$$\begin{array}{l} CH_2OC(O)R_1 \\ | \\ CHOC(O)R_2 \\ | \\ CH_2OC(O)R_3 \end{array}$$

where $R_1$, $R_2$, and $R_3$ are saturated or unsaturated hydrocarbon chains of various lengths. When relatively short chains predominate, the fat is an oil; with chains of greater length, it is a solid. These glycerides are too reactive to have been preserved in fossils, even of Pleistocene age. As man long ago discovered, in the presence of alkali they break down into glycerin and soaps that are salts of the acids based on $R_1$, $R_2$, and $R_3$. Presumably fatty acids found in fossil specimens and fossiliferous rocks had their origin in these glycerides.

Waxes are another type of lipid produced by both animals and plants. They are esters of a fatty acid and a long-chain alcohol. Cetyl palmitate, $C_{15}H_{31}COOC_{16}H_{33}$, found in whales and certain invertebrates, is typical. Other waxy esters form part of the glistening coats on many leaves and fruits. All are very insoluble in water and no more than slightly soluble in the organic solvents that dissolve the fatty acids. Like glycerides, they split in the presence of alkali, but the resulting alcohols are insoluble enough to be preserved in fossils together with fatty acid and undecomposed wax. Other lipoidal substances having sufficient chemical stability to exist in fossils are hydrocarbons of high molecular weight, a wide range of branched-chain isoprene derivatives including squalene, and the carotenoids and heterocyclic terpenoids such as cholesterol and other sterols. These substances occur in petroleum, in soils, and in recent sediments (reviewed in Breger, 1963; *see also* Kvenvolden, 1967). Esters of cholesterol and fatty acids have been identified in some well preserved Pleistocene specimens (Das, Doberenz, and Wyckoff, 1967); otherwise only free fatty acids and an occasional phospholipid have been sought and found among the individual animal fossils that might still contain them.

This initial concentration of attention on the occurrence of fatty acids is to be expected in view of their relative importance and the ease with which they can be investigated by gas chromatography. They occur in nature as complex mixtures of components having very similar chemical properties, and the development of chromatography has incredibly simplified their precise analysis.

The instrument is that used in the analysis of amino-acids. The temperature programming there required can be dispensed with, however, and since fatty acids do not react with stainless steel, the column of the chromatograph need not be of glass. The packing material for the column varies with the substances being analysed, but in the study of fossils diethyleneglycol succinate on Chromosorb has been found satisfactory.

The following simple procedure (Everts, Doberenz, and Wyckoff, 1968) will isolate the fatty acids in a fossil bone or tooth. Carbonates are removed from the powdered sample by a preliminary treatment with HCl till evolution of $CO_2$ ceases. If the undissolved bone is then immediately separated by centrifugation, little if any lipoid is lost in the discarded supernatant. Fatty acids may be freed from the treated bone and at the same time converted to their methyl esters by refluxing the dried residue overnight in $2N$ methanolic HCl followed by extraction of the solution with $CS_2$ and subsequent evaporation of the $CS_2$ layer. If a sufficient amount is present, the esters can be weighed in a microbalance before being redissolved in $CS_2$ to furnish the sample for chromatography. When other lipids are suspected, the fossil should be extracted with a chloroform–methyl alcohol (2 : 1) mixture and the residue refluxed with methylated $2N$ HCl after carbonate removal. As with other gas chromatograms, areas under the peaks in the record are proportional to the amounts of the responsible fatty acid esters; they are identified and absolute quantities ascertained by comparison with runs on standard mixtures of the pure esters.

Unsaturated fatty acids are in general less stable than their saturated analogues and it has therefore been rather surprising to find appreciable quantities in a number of fossils. Their presence can be verified by the familiar iodine test but it is best done by hydrogenation. Their amounts can be established by analyses before and after this conversion to their saturated analogues. Various hydrogenation procedures are known, but with fossils it is easily carried out (Farquhar, Insul, Rosen, Stoffel, and Ahrens, 1959) by reacting an alcoholic solution of the fatty acids before methylation with gaseous hydrogen in the presence of platinum oxide as catalyst.

## *Table XXXVIII.*—Typical Fatty Acid Analyses of Recent and Fossil Bones

| Fossil | Acid: Relative Peak Areas (Percentage of Total) | | | | | | | | | Amount (μg. per g.) |
|---|---|---|---|---|---|---|---|---|---|---|
| | 10* | 12 | 14 | 16 | 16:1 | 18 | 18:1 | 20 | Unidentified | |
| **Recent** | | | | | | | | | | |
| V63689 | Trace | Trace | 3·4 | 24·2 | 9·7 | 9·7 | 47·6 | 0 | 1·7 | 19·6 |
| V6788 | Trace | 0·6 | 2·7 | 30·7 | 2·2 | 8·2 | 43·2 | 0 | 0·3 | — |
| **Pleistocene** | | | | | | | | | | |
| V64118 | 0 | Trace | 10·4 | 32·6 | 24·3 | 11·2 | Trace | 0 | 19·5 | 0·64 |
| V63447 | 0 | Trace | 19·4 | 42·0 | 13·6 | 20·2 | 0 | 0 | 4·8 | 0·26 |
| V63234 | Trace | Trace | 3·5 | 45·6 | 17·9 | 23·9 | Trace | 0 | 0·1 | 0·76 |
| **Pliocene** | | | | | | | | | | |
| V63233 | 0 | 9·3 | 12·5 | 28·8 | 10·7 | 13·1 | 4·2 | 0 | 21·4 | 1·02 |
| V63509 | Trace | 4·2 | 8·7 | 19·5 | 30·6 | 12·5 | 3·7 | 0 | 20·8 | 0·48 |
| V63539 | 0 | 6·6 | 11·8 | 26·3 | 16·9 | 12·1 | 0 | 0 | 26·3 | 0·62 |
| **Miocene** | | | | | | | | | | |
| V67116 | Trace | 8·9 | 8·0 | 42·6 | 6·9 | 12·7 | Trace | 18·3 | 2·6 | 0·62 |
| V64117 | 8·5 | 17·5 | 11·8 | 13·2 | 4·6 | Trace | 7·9 | 14·8 | 15·0 | 0·88 |
| **Oligocene** | | | | | | | | | | |
| V6416 | 0 | 11·4 | 11·4 | 42·8 | 4·6 | 16·3 | 0 | 7·8 | 5·6 | — |
| V6835 | 0 | 2·5 | 8·8 | 28·3 | 32·9 | 1·8 | 8·9 | 13·3 | 1·5 | — |
| **Eocene** | | | | | | | | | | |
| V6383 | 1·2 | 1·8 | 3·1 | 14·7 | 13·0 | 11·7 | 0 | 14·1 | 24·6 | 1·09 |
| V64113 | 1·4 | 1·7 | 28·3 | 37·1 | 13·3 | 6·8 | 1·4 | Trace | 9·6 | — |
| **Cretaceous** | | | | | | | | | | |
| V6613 | 2·0 | 3·9 | 7·9 | 27·7 | 29·5 | 1·5 | 6·5 | 9·6 | 2·2 | 0·68 |
| V666 | 1·7 | 4·7 | 13·0 | 26·8 | 23·5 | 1·7 | 5·1 | 0 | 22·3 | 1·10 |
| V67155 | Trace | 2·6 | 5·4 | 20·7 | 8·7 | 9·3 | 7·9 | 0 | 30·5 | 0·68 |
| **Jurassic** | | | | | | | | | | |
| V674 | 0 | 7·3 | 8·3 | 23·1 | 6·7 | 18·8 | 24·6 | 0 | 11·2 | 0·85 |
| V67138 | 10·2 | 5·5 | 4·9 | 30·7 | 27·1 | 0 | 14·4 | 0 | 7·2 | 0·52 |
| V67148 | 6·0 | 4·9 | 8·9 | 35·0 | 29·3 | Trace | 10·6 | 0 | 5·3 | — |
| **Triassic** | | | | | | | | | | |
| V6857 | 1·4 | 6·3 | 16·8 | 39·5 | 18·6 | 3·1 | Trace | 9·0 | 5·3 | 0·45 |
| V6610 | 0 | 22·4 | 13·3 | 30·0 | 20·7 | 9·8 | Trace | 0 | 3·7 | — |
| V671 | 0 | 12·7 | 10·0 | 24·4 | 0 | 31·0 | 0 | 0 | 15·6 | 0·60 |
| **Permian** | | | | | | | | | | |
| V664 | 1·2 | 8·9 | 6·6 | 13·7 | 35·1 | 0·9 | 6·2 | 0 | 27·4 | 1·00 |
| V631 | 1·5 | 9·1 | 7·8 | 18·8 | 33·7 | 9·5 | 2·0 | 0 | 12·4 | 0·92 |
| **Pennsylvanian** | | | | | | | | | | |
| V63724 | 0 | 10·2 | 6·5 | 30·7 | 25·3 | 8·2 | 10·7 | 3·8 | 4·5 | — |

* The numbers in this line identify the fatty acids in terms of the number of carbon atoms in their molecules. Thus stearic acid is designated as 18 and palmitic as 16. An unsaturated acid is indicated by an added colon and the number of its double bonds (e.g., oleic acid = 18:1).

When their glycerides are broken down, the same broad spectrum of fatty acids is obtained from all sorts of fresh bones and teeth, though some variations are noted in relative amounts depending on the type of animal. The unsaturated oleic acid [$CH_3(CH_2)_7CH=CH(CH_2)_7COOH$] is always the most prominent component. Next most important are members of the saturated straight-chain series containing an even number of carbon atoms together with certain of their unsaturated analogues. In decreasing order of amounts usually obtained from fresh mammalian bones the acids are:

palmitic [$CH_3(CH_2)_{14}COOH$],
stearic [$CH_3(CH_2)_{16}COOH$],
linoleic [$CH_3(CH_2)_4CH=CHCH_2CH=CH(CH_2)_7COOH$],
palmitoleic [$CH_3(CH_2)_5CH=CH(CH_2)_7COOH$] and
myristic [$CH_3(CH_2)_{12}COOH$].

The oils and waxes accumulated by marine vertebrates such as whales and sharks and by some corals and other invertebrates do not yield this pattern of fatty acids.

Little oleic acid is retained in fossil bones and teeth but some palmitoleic acid is not infrequently found (Everts and others, 1968; Everts, 1969). In other respects the spectrum of their fatty acids resembles that of fresh specimens. Averages from a considerable number of fossil bones, some as old as the Pennsylvanian, have shown the decreasing sequence: palmitic, stearic, palmitoleic, myristic, lauric [$CH_3(CH_2)_{10}COOH$], and oleic acids. Typical analyses from this series are shown in *Tables XXXVIII* and *XXXIX*. They illustrate several significant facts. In the first place there are surprisingly small differences in content for the analysed fossils, and the amount of recoverable fatty acids does not progressively diminish with age. Variations in amount must rather be attributed to conditions of fossilization and preservation. The data are still too few to show with any certainty if there are species-dependent differences in the ratios of the several acids. On the whole, the fossils contain a relatively higher proportion of the shorter-chain myristic and lauric acids than do fresh bones; it would, however, be premature to decide whether this is characteristic of older forms of life or a consequence of the slow fragmentation of long-chain compounds that occurs under certain conditions. Another mark of the incompleteness of present knowledge is the number of still unidentified peaks which demonstrably are not due to members of the straight-chain series of acids.

When seeking to ascertain the source of the fatty acids in individual fossils, one encounters questions already presented by their proteins and amino-acids: Are the acids native to the fossil or contaminants that have entered from the surrounding rock or soil? Fatty acids are not, like amino-acids, readily soluble in

*Table XXXIX.*—TYPICAL ANALYTIC DATA ON UNSATURATED ACIDS IN FOSSILS

| FOSSIL | RELATIVE PEAK AREAS (PERCENTAGE OF TOTAL) | | | |
|---|---|---|---|---|
| | Before Hydrogenation | | After Hydrogenation | |
| | 16 | 16:1 | 16 | 16:1 |
| Pleistocene | | | | |
| V64118 | 32·6 | 24·3 | 51·7 | 5·2 |
| V63421 | 31·1 | 18·5 | 36·8 | 12·8 |
| Pliocene | | | | |
| V6528 | 20·3 | 21·7 | 21·0 | 21·0 |
| V63509 | 19·5 | 30·6 | 36·2 | 13·9 |
| Eocene | | | | |
| V6383 | 14·7 | 13·0 | 19·3 | 8·4 |
| Permian | | | | |
| V631 | 18·8 | 33·7 | 40·4 | 12·1 |
| V662 | 24·1 | 20·8 | 36·2 | 8·5 |

water, but their limited solubilities give them a slow mobility that cannot be ignored. They are present in soil and clays and could enter from them into fossils lying in loose deposits. When they are absent from the denser rocks that enclose fat-containing fossils, it is difficult to avoid the conclusion that they are native to the specimen. It is harder to decide if they had their origin in the organic tissue of the fossil or are products of post-mortem decomposition and bacterial action. Bacteria contain very long-chain fatty acids and ones incorporating cyclopropane groups in addition to those found in fresh bones. When such acids are present in a fossil, serious consideration must be given to the possibility of a bacterial origin.

In contrast to the relatively meagre attention given lipids in fossils, much work has been devoted to their occurrence in soils, in marine sediments, and in crude petroleum. In all of these, fatty substances are found in great profusion. The results obtained lie outside the province of this monograph though there is an ultimate connexion between what is found in individual fossils, fossiliferous rocks, and ancient sediments. The last two may contain kerogen, which yields long-chain hydrocarbons on heating and long-chain acids through oxidation with chromic acid (Hoering and Abelson, 1965). The acids produced depend, however, on the way oxidation is carried out and consequently there is no direct relation between them and the lipids that presumably contributed to the formation of the kerogen.

For the student of the lipids in individual fossils the two most important tasks awaiting attention clearly are the analysis of specimens of great age and the identification of the substances responsible for the numerous unknown chromatographic peaks given by fossils of all ages. Until more progress has been made in these directions it will be hard to evaluate the contribution that knowledge of the fatty-acid content can make to our understanding of early forms of life.

## CHAPTER VI

# SUMMARY

THE results discussed in the preceding chapters are as yet quite insufficient to answer most of the questions raised in the Introduction. Many more specimens must be analysed and an intensified effort must be made to include those in the best possible state of preservation; much early work has of necessity been done on rather mediocre material.

As indicated in CHAPTER II, the original details of structure of many fossils have remained practically unaltered for hundreds of millions of years. The development of electron optical instruments has provided the tools needed to establish this fact down to the level of the macromolecules that are the structural units of living matter. Such examination is supplying an invaluable means of selecting specimens most worthy of intensive study and the objective basis for interpretations arising out of it. The practically perfect preservation of structure it often reveals furthermore shows the feasibility of seeking to develop a comparative micromorphology that will trace at this molecular level the way bones, teeth, and shells have developed during the course of evolution.

The results outlined in CHAPTER IV suggest how the discovery that proteinaceous residues persist in fossils of great age opens up promising avenues for further investigation. The presence of amino-acids, as products of protein decomposition, has been known for some time, but much more can be learned from the insoluble residues now being recovered. They are more or less complete proteins in the sense that they are composed of most of the amino-acids found in living matter. In many well-preserved bones and teeth their composition is close to that of Modern collagen; in other specimens it is very different. Pleistocene fossils may contain about as much of this protein as fresh bones; enough for complete analyses has been isolated from numerous specimens more than 100 million years old. Though invertebrate shells rather quickly lose a large part of their original protein, the oldest yet examined have retained sufficient quantities for study. It is a noteworthy fact that after an initial fall the amounts retained in both vertebrate and invertebrate fossils do not show a further steady decline with age; clearly these residues are so stable that they may confidently be sought in the oldest obtainable specimens.

The results to date suggest several directions for future inquiry. In view of the unexpectedly good preservation of collagen in many Pleistocene fossils and the success achieved years ago in blood-typing human mummies, a study of the immunological characteristics of mummified soft tissues should be especially rewarding. Analyses made of the oldest fossils thus far studied do not suggest that their proteins were chemically any simpler than those now being produced. Modern proteins differ from one another in the ways their amino-acids are distributed within the molecule, and these characteristic details of molecular architecture are being established by the newly developed techniques of a sequential analysis which shows how the acids succeed one another in the peptides into which a protein molecule is split. We are now in a position to begin similar studies of fossil proteinaceous residues in the search for evolutionary progressions in the organisms themselves.

The first step in interpreting a residue must be a determination of whether it is contemporaneous with the fossil or a more recent contaminant entering from the soil, from micro-organisms growing on it, or from handling during examination and analysis. The factors underlying this determination are discussed in CHAPTER IV. Residues proper to the fossil could have diverse origins. Some clearly derive from the original tissue proteins. Nevertheless the proteins of most uneaten carcasses are replaced by those of scavenging organisms of one sort or another. It is therefore not surprising that certain fossil proteins have compositions very like those of bacteria.

Though analyses recorded here demonstrate that proteins can endure for geological periods, even the most stable undoubtedly alter with time. This process of slow denaturation must be occurring in fossils. Like most chemical reactions it is greatly accelerated by heat and therefore can be studied in the laboratory.

Experiments to this end have been begun on both collagen and the conchiolins of shells. Moist collagen readily converts to soluble gelatin, but when thoroughly dry it disappears much more slowly and with only a gradual shift in composition. Unaltered collagen as a consequence will only be found in fossil bones and teeth that have never been exposed to moisture at temperatures much above normal. Studies of the thermal stabilities of the proteins in fresh shells have shown that the residues after heating are often similar to those isolated from fossils. Composition depends on the type of shell, but for related species it is nearly the same for heated fresh shells and for their fossils older than Pliocene. In such cases it would appear that the fossil proteins have arisen through slow denaturation of one or more components of the original conchiolin. Further analyses should make definite the characteristic relationships that undoubtedly exist between the fossil proteins of different types of invertebrate.

Valuable information about the past history of a fossil can be gained through investigation of the free amino-acids and peptides present as fragments of its disintegrated protein. The more stable amino-acids will persist and will be supplemented by others arising through the transformation of some that have disappeared. Among the newly formed acids are isomers of the original acids and compounds that do not enter into the composition of known proteins. The rates of some of these transformations are known and can be used to estimate age and other features of a fossil's past history; much more can, however, be learned when accurate measurements have been made of the response of each amino-acid to heat.

Still more about the past of a fossil can, as indicated in CHAPTER III, be ascertained through the close study of its inorganic framework. Thus whether or not an initially aragonitic shell has become calcitic will depend on the temperature and amount of moisture to which it has been exposed. Analyses for magnesium and strontium under certain circumstances furnish an index of the temperature of the water in which the animal grew. Other chemical elements are found in trace amounts. Often they have been derived from surrounding rocks but in some instances they probably entered the bone, tooth, or shell before fossilization. Then they may give information about conditions under which the animal lived, and perhaps died.

Fatty acids are the only other components of animal fossils yet to have received serious attention. They are found in specimens of all ages (CHAPTER V). For the most part the fatty acids in fossils are also present in living bone, but the chromatograms from some fossils have had important peaks of unknown origin. Until these have been identified it is impossible to know how rewarding will be a more intensive study.

The investigations thus far made of the microstructure and composition of individual fossils have an exploratory character that is gradually showing how to select the most fruitful directions for future effort. Some of these will concentrate on questions discussed in this monograph, others will be aimed at problems that will only emerge as exploration proceeds. The results to date are sufficient to justify this further exploration. Their chief importance, however, lies in the demonstration that the microscopic and analytical tools developed in recent years provide a sound basis for a direct investigation of the relatively unaltered remains of earlier life which they have already shown to exist.

# *APPENDIX* I

## GEOLOGIC TIME SCALE

| Period | Epoch | Beginning (Years ago) |
|---|---|---|
| | *Cenozoic Era* | |
| Quaternary | Pleistocene | About 30,000 |
| Tertiary | Pliocene | About 12 million |
| | Miocene | 25 million |
| | Oligocene | 35 million |
| | Eocene | 60 million |
| | Paleocene | 65 million |
| | *Mesozoic Era* | |
| Cretaceous | Upper | 70 million |
| | Lower | 110 million |
| Jurassic | Upper | 160 million |
| | Lower | 180 million |
| Triassic | Upper | 200 million |
| | Lower | 230 million |
| | *Palaeozoic Era* | |
| Permian | Upper | 260 million |
| | Lower | 280 million |
| Pennsylvanian | Carboniferous | 320 million |
| Mississippian | | 340 million |
| Devonian | Upper | 360 million |
| | Lower | 400 million |
| Silurian | | 420 million |
| Ordovician | Upper | 440 million |
| | Lower | 500 million |
| Cambrian | | 530 million |

*APPENDIX 2*

## SPECIMEN IDENTIFICATION

THIS appendix lists the fossil specimens to which reference is made in the text. It states, after this laboratory's identifying numbers, only enough information to make intelligible their use in text, tables, and figure legends. In most cases common names have been employed. Where these fossils have formed part of published work further details can be found in the appropriate articles.

Many of the specimens have been collected from nearby localities by members of the laboratory staff. Other persons have also contributed specimens that have been invaluable in furthering the work. We especially wish to express gratitude to the following: J. R. Macdonald and the Los Angeles County Museum of Natural History; J. A. Jensen and the Earth Sciences Museum of Brigham Young University; E. H. Colbert and the American Museum of Natural History; R. Dughi and the Musée d'Histoire Naturelle d'Aix-en-Provence; M. Dawson and the Carnegie Museum (Pittsburgh).

V631: *Bradysaurus* (primitive reptile) bone fragment, Permian.
V633: *Desmostylus* tooth fragment, Miocene, Monocline Ridge, California.
V6351: *Plesippus* (early horse) tooth, Plio-Pleistocene, City Canyon, Texas.
V6383: Turtle scute fragments, Eocene, Tabernacle Butte, Wyoming.
V63101: Dicynodont (reptilian) bone, Triassic, Karroo Beds, South Africa.
V63184: *Stegomastodon* molar, Pleistocene, Safford, Arizona.
V63210: *Equus* (horse) bone, Pleistocene, Safford, Arizona.
V63217: Horse molar, Pleistocene, Rancho La Brea, California.
V63218: Bison tooth, Pleistocene, Rancho La Brea, California.
V63220: *Panthera atrox* tooth, Pleistocene, Rancho La Brea, California.
V63233: Horse bone, Pliocene, Wikieup, Arizona.
V63234: Mastodon bone, Pleistocene, Benson, Arizona.

V63244: Ungulate bone, Pleistocene, Rancho La Brea, California.
V63247: *Canis dirus* (wolf) bone, Pleistocene, Rancho La Brea, California.
V63289: Camel bone, Pleistocene, Benson, Arizona.
V63291: Ungulate bone, Pleistocene, Benson, Arizona.
V63295: Ungulate bone, Pleistocene, Benson, Arizona.
V63378: Mastodon bone, Pleistocene, Benson, Arizona.
V63421: Unidentified bone scrap, Pleistocene, Benson, Arizona.
V63447: Unidentified bone scrap, Pleistocene, Benson, Arizona.
V63456: Mastodon tusk, Pleistocene, Benson, Arizona.
V63477: Ungulate bone scrap, Pliocene, Wikieup, Arizona.
V63481: Ungulate bone scrap, Pliocene, Wikieup, Arizona.
V63486: Ungulate bone scrap, Pliocene, Wikieup, Arizona.
V63495: Ungulate bone scrap, Pliocene, Wikieup, Arizona.
V63509: Ungulate bone scrap, Pliocene, Wikieup, Arizona.
V63526: Ungulate bone scrap, Pliocene, Wikieup, Arizona.
V63539: Ungulate bone scrap, Pliocene, Wikieup, Arizona.
V63562: Ungulate bone scrap, Pliocene, Walnut Grove, Arizona.
V63636: Phytosaur tooth, Triassic, Petrified Forest, Arizona.
V63689: Bones, cow, Recent, Arizona desert, sun-dried.
I63693: *Venus campechiensis*, Pleistocene, Cape May, New Jersey.
V63724: *Sagenodus copeanus* (lung fish) bone, Pennsylvanian, Kansas.
V642: Horse bone, Pleistocene, Rancho La Brea, California.
V648: Ground sloth tooth, Pleistocene, Rancho La Brea, California.
V6410: Ungulate tooth, Pleistocene, Egypt.
V6416: Mammal bone, Oligocene, Egypt.
V6417: *Brontotherium* tooth, Oligocene, White River Farm, Nebraska.
V64113: *Pristis* (sawfish, ray) bone fragment, Eocene, Egypt.
V64117: Sirenian rib, Miocene, California.
V64118: Mammoth tusk, Pleistocene, Irvington, California.
V64142: *Pristis* (sawfish, ray) bone fragment, Eocene, Egypt.
V64145: *Felis atrox* tooth, Pleistocene, Rancho La Brea, California.
V64148: *Smilodon* (*californicus*) bone, Pleistocene, Rancho La Brea, California.
V64149: *Smilodon* bone, Pleistocene, Rancho La Brea, California.
V64151: Ground sloth tooth, Miocene, Colombia.
V64184: Gopher bone, Pleistocene, Rancho La Brea, California.
V64185: Gopher jaw bone, Pleistocene, Rancho La Brea, California.
V655: Camel bone, Pleistocene, Rancho La Brea, California.
V656: *Smilodon* bone, Pleistocene, Rancho La Brea, California.
V657: Bison bone, Pleistocene, Rancho La Brea, California.
V658: Ground sloth bone, Pleistocene, Rancho La Brea, California.
V659: Camel jawbone, Pleistocene, Rancho La Brea, California.
V6511: *Canis dirus* (wolf) jawbone, Pleistocene, Rancho La Brea, California.
V6512: Coyote jawbone, Pleistocene, Rancho La Brea, California.
V6513: *Panthera atrox* jawbone, Pleistocene, Rancho La Brea, California.
V6515: Horse bone, Pleistocene, Rancho La Brea, California.

V6519: Dinosaur bone, Cretaceous, Texas.
V6521: Bear bone, Pleistocene, Rancho La Brea, California.
V6522: *Panthera atrox* bone, Pleistocene, Rancho La Brea, California.
V6524: Mastodon tooth, Pleistocene, Rancho La Brea, California.
V6526: Bison bone, Pleistocene, Rancho La Brea, California.
V6528: Horse bone, Pliocene, Hagerman, Idaho.
V662: *Sphenacodon* (Pelycosaur reptile) bone, Permian, New Mexico.
V664: *Diadectes* (primitive reptile) bone, Permian, Seymour, Texas.
V666: Mosasaur (marine lizard) bone, Cretaceous, Niabraro Formation (Fm), Kansas.
V6610: Therapsid reptilian bone, Triassic, Chinle Fm, Arizona.
V6613: Mosasaur (marine lizard) vertebra, Cretaceous, Panoche Hills, California.
V6643: Mammoth tooth, Pleistocene, Tuba City, Arizona.
V671: Phytosaur skull bones, Triassic, Chinle Fm, Arizona.
V674: Sauropod dinosaur bones, Jurassic, Morrison Fm, Wyoming.
V675: Hadrosaur (duck billed) dinosaur bones, Cretaceous, Oldman Fm, Alberta, Canada.
V6721: Bison bone, Hell Gap, Wyoming.
V6731: Mammoth bone, Early man, Union Pacific Mammoth Site, Wyoming.
V6733: Mammoth tusk, Early man, Union Pacific Mammoth Site, Wyoming.
V6734: Bison bone, Early man, Olsen-Chubbuck, Colorado.
V6783: Horse teeth, Pliocene, Wikieup, Arizona.
V6788: Human femur, Modern.
I67114: *Pecten madisonius*, Miocene, Jamestown Ferry, Virginia.
V67116: Whale bones, Miocene, Jamestown Ferry, Virginia.
V67138: Sauropod dinosaur limb bone, Jurassic, Morrison Fm, Wyoming.
V67140: Dinosaur bone scrap, Cretaceous, Lance Fm, Montana.
V67142: Dinosaur bone scrap, Cretaceous, Edmonton Fm, Montana.
V67146: Dinosaur toe bone, Cretaceous, Lance Fm, Montana.
V67147: Dinosaur toe bone, Cretaceous, Lance Fm, Montana.
V67148: Dinosaur sacrum, Jurassic, Morrison Fm, Wyoming.
V67150: *Diplodocus* dinosaur bones, Jurassic, Morrison Fm, Wyoming.
V67154: Mosasaur (marine lizard) bone scrap, Cretaceous, Niobrara Fm, Kansas.
V67155: Mosasaur (marine lizard) bone scrap, Cretaceous, Niobrara Fm, Kansas.
V67156: Crocodilian bone, Cretaceous, Greensand Fm, New Jersey.
V67162: Mosasaur (marine lizard) bones, Cretaceous, Niobrara Fm, Kansas.
V6816: Diceratherium (rhinoceros) tooth, Miocene, Agate Springs, Nebraska.
V6820: *Merycodus* (prongbuck) bones, Plicene, Valentine Fm, Nebraska.
V6828: Teleost fish skeleton, Jurassic, Antofagasta Province, Chile.

V6835: Tortoise carapace, Oligocene, Brule Fm, South Dakota.
V6837: Hadrosaur (duck bill dinosaur) tendon, Cretaceous, Hell Creek Fm, Montana.
V6843: Dinosaur (?) egg-shells, Cretaceous, North Horn Fm, Utah.
V6844: Dinosaur (?) egg-shells, Cretaceous, Cedar Mountain Fm, Utah.
V6846: *Aepyornis* egg-shell, Pleistocene, Madagascar.
V6850: Garfish scales, Eocene, locality unknown.
V6851: Dinosaur toe bone, Cretaceous, Lance Fm, Wyoming.
V6852: Dinosaur bone fragment, Jurassic, Morrison Fm, Wyoming.
V6857: *Eupelor* (amphibian) vertebra, Triassic, Chinle Fm, Arizona.
V6860: Dinosaur tooth, Cretaceous.
V6867: Crocodilian bone, Cretaceous, Greensand Fm, New Jersey.
V6869: Dinosaur bones, Cretaceous, Laramie Beds, Montana.
V6877: Redwinged blackbird egg, Recent.
V6878: Great blue heron egg-shells, Recent.
V6879: Black-necked stilt egg-shells, Recent.
V6880: Magpie egg-shells, Recent.
V6884: Human tooth, about 2000 years old.
V6885: *Basiliscus plumbifrons* (basilisk) egg-shell, Recent.
V6887: Rhinoceros iguana egg-shells, Recent.
V6888: *Gopherus agassizi* (desert tortoise) egg-shells, Recent.
V6889: *Testudo elephantopus* (Galapagos tortoise) egg-shells, Recent.
V6890: *Crocodylus johnsoni* egg-shell, Recent.
V6891: *Malacocherus tornieri* (African soft-shelled tortoise) egg-shells, Recent.
V6893: *Alligator mississippiensis* egg-shells, Recent.
V6894: *Dromiceius novae hollandiae* (emu) egg-shells, Recent, Australia.
V6896: *Casuarius unappendiculatus* (cassowary) egg-shells, Recent, Island of Salawatti.
V6897: *Struthio camelus australis* (ostrich) egg-shells, Recent, South Africa.
V6898: *Alligator mississippiensis* egg-shells, Recent, Florida.
V68101: Dinosaur egg-shells, Cretaceous, Gobi Desert.
V68108: Ungulate tooth fragments, Pliocene, Wikieup, Arizona.
V68112: *Gallus gallus* (chicken) egg-shells, Recent, Tucson, Arizona.
V68113: Horse tooth fragments, Pliocene, Wikieup, Arizona.
V696: *Ptychodus* (skate) tooth, Cretaceous, South Dakota.
V697: *Taeniolabis* (primitive mammal) teeth, Palaeocene, Patterson Site, Wyoming.
V6912: Rhinoceros teeth, Oligocene, Brule Fm, Wyoming.
V6931: *Aepyornis* egg-shells, Pleistocene, Madagascar.
I6932: *Crassostrea virginica*, Pleistocene, Horry County, South Carolina.
I6933: *Ostrea sculpturata*, Pliocene, Myrtle Beach, South Carolina.
I6935: *Ostrea ventilabum*, Oligocene, Klein-Spauwen, Belgium.
I6937: *Ostrea johnsoni*, Eocene, Jackson, Clarke County, Alabama.
I6938: *Ostrea compressirostra*, Palaeocene, Potomac Creek, Virginia.

I6941: *Gryphaea convexa*, Cretaceous, Cream Ridge, New Jersey.
V6946: *Ornitholithus arcuatus* egg-shells, Eocene, Bassin-du-Rhône, St. Antonin-sur-Bayon, France.
I6954: *Ostrea* sp. (oysters), Miocene, Calvert Beach, Maryland.
I6959: *Pecten eboreus*, Pliocene, Horry County, South Carolina.
I6960: *Pecten crassicardo*, Miocene, Cammati Canyon, California.
I6961: *Chlamys madisonius*, Miocene, Calvert Beach, Maryland.
I6962: *Glycymeris americana*, Pliocene, Myrtle Beach, South Carolina.
I6966: *Cardita planicosta*, Eocene, Bracklesham Fm, England.
I6989: *Ostrea* sp., Pleistocene, Huatabampito, Sinaloa, Mexico.
I6990: *Arca pacifica*, Pleistocene, Huatabampito, Sinaloa, Mexico.
I6994: *Dinocardium robustum*, Pleistocene, Horry County, South Carolina.
I6995: *Chione morsitans*, Pleistocene, Horry County, South Carolina.
I6996: *Saxidomus giganteus*, Pleistocene, San Mateo County, California.
I6997: *Mercenaria carolinensis*, Pleistocene, Horry County, South Carolina.
I6998: *Mercenaria* sp., Pleistocene, Wailes Bluff, Maryland.
I6999: *Dosinia discus*, Pleistocene, Virginia Beach, Virginia.
I69100: *Lunatia* sp. (gastropod), Pleistocene, Virginia Beach, Virginia.
I69101: *Busycon carica* (gastropod), Pleistocene, New Bern, North Carolina.
I709: *Pseudopecten equivalvis*, Jurassic, Blockley, Gloucestershire, England.
I7018: Corals (coelenterates), Pleistocene, Fort Myers, Florida.
I7021: *Diploria strigosa* (coelenterate), Pleistocene, Miami oölite, Florida Keys.
I7022: *Siderastrea* sp. (coelenterate), Pleistocene, Miami oölite, Florida Keys.
I7027: *Pecten* sp., Pleistocene, Florida.
I7029: *Aequipecten circularis*, Pliocene, Kettleman Hills, California.
I7030: *Pecten bellus*, Pliocene, Santa Barbara, California.
I7031: *Chlamys eboreus*, Pliocene, Myrtle Beach, South Carolina.
I7032: *Pecten byramensis*, Oligocene, Vicksburg, Mississippi.
I7033: *Pecten texanus*, Cretaceous, Tarrant County, Texas.
V7034: Ornithopod dinosaur egg-shells, Cretaceous, Bassin d'Aix-en-Provence, Roques Hautes, France.
V7035: Sauropod dinosaur egg-shells, Cretaceous, Bassin d'Aix-en-Provence, Trets, France.
V7036: Theropod dinosaur egg-shells, Cretaceous, Bassin d'Aix-en-Provence, Rousset, France.
I7046: *Argopecten alquisulcatus*, Pleistocene, San Pedro, California.
I7048: *Pecten bellus*, Pleistocene, Coyote Hills, California.
I7049: *Pecten eldridgei*, Pleistocene, McKittrick County, California.
I7050: *Pecten latiauratus*, Pleistocene, Playa del Rey, California.
I7074: *Arca pacifica*, Recent, San Felipe, Baja California, Mexico.
I7080: *Chione gnidia*, Pleistocene, Huatabampito, Sinaloa, Mexico.
I711: *Pinctada* sp., Recent, Japan.

# BIBLIOGRAPHY

ABDERHALDEN, E., and HEYNS, K. (1933), 'Nachweis von Chitin in Flügelresten von Coleopteren des oberen Mitteleocäns (Fundstelle Geiseltal)', *Biochem. Z.*, **259**, 320–321.

ABELSON, P. H. (1954), 'Organic constituents of fossils', *Carnegie Inst. Wash. Year Book*, **53**, 97–101.

ABELSON, P. H. (1955), 'Organic constituents of fossils', *Carnegie Inst. Wash. Year Book*, **54**, 107–109.

ABELSON, P. H. (1956), 'Paleobiochemistry', *Sci. American*, **195**, 83–92.

ABELSON, P. H. (1957), 'Some aspects of paleobiochemistry', *Ann. N.Y. Acad. Sci.*, **69**, 276–285.

ABELSON, P. H. (1963), 'Geochemistry of amino acids', in *Organic Geochemistry* (ed. BREGER, I. A.), Monograph No. 16 Earth Science Series, pp. 431–455. Oxford: Pergamon. Distributed by the Macmillan Co., New York.

ABELSON, P. H., HOERING, T. C., and PARKER, P. L. (1963), 'Fatty acids in sedimentary rocks', in *Advances in Org. Geochem.*, Proc. Int. Meeting, Milan, 1962, pp. 169–174. New York: Pergamon.

AKIYAMA, M. (1964), 'Quantitative analysis of the amino acids included in Japanese fossil scallop shells', *J. geol. Soc. Japan*, **70**, 508–516.

AKIYAMA, M. (1966), 'Conchiolin-constituent amino acids and shell structures of bivalved shells', *Proc. Japan Acad.*, **42**, 800–805.

AKIYAMA, M. (1971), 'Amino acid composition of fossil scallop shell proteins and non-proteins', *Biomineralization*, **3**, 65–70.

AKIYAMA M., DAVIDSON, F. D., MATTER, P., III, and WYCKOFF, R. W. G. (1971), 'Liquid and gas chromatographic analyses of several fossil proteins', *Comp. Biochem. Physiol.*, **38B**, 93–101.

AKIYAMA, M., and FUJIWARA, T. (1966), 'Amino acid survivals of fossil shells yielded from the Johmon shell mounds in the Kanto district, Japan', *Misc. Repts, Res. Inst. Nat. Resour.* (*Tokyo*), No. 67, 67–72.

AKIYAMA, M., FUJIWARA, T., HOTTA, S., KAIBARA, H., KAMEI, T., KOBAYASHI, I., and SATO, T. (1968), 'Histological and biochemical studies on fossil elephant teeth of *Elephas naumanni* Makiyama', *Kaseki Kenkyu Kaishi*, **1**, 35–75.

AKIYAMA, M., MATTER, P., III, and WYCKOFF, R. W. G. (1970), unpublished data.

AKIYAMA, M., and WYCKOFF, R. W. G. (1970), 'Total amino acid content of fossil pecten shells', *Proc. Natn. Acad. Sci. U.S.A.*, **67**, 1097–1100.

ARMSTRONG, W. G., and TARLO, L. B. H. (1966), 'Amino acid components in fossil calcified tissues', *Nature, Lond.*, **210**, 481–482.

ASCENZI, A. (1955), 'Some histochemical properties of the organic substance in Neandertalian bone', *Am. J. phys. Anthrop.*, N.S., **13**, 557–566.

BARGHOORN, E. S., MEINSCHEIN, W. G., and SCHOPF, J. W. (1965), 'Paleobiology of a Precambrian shale', *Science, N.Y.*, **148**, 461–472.

BARGHOORN, E. S., and SCHOPF, J. W. (1965), 'Microorganisms from the Late Precambrian of Central Australia', *Science, N.Y.*, **150**, 337–339.

BARGHOORN, E. S., and SCHOPF, J. W. (1966), 'Microorganisms three billion years old from the Precambrian of S. Africa', *Science, N.Y.*, **152**, 758–763.

BAUD, C.-A., and MORGENTHALER, P. W. (1952), 'Recherches sur l'ultrastructure de l'os humain fossile', *Archs suisses Anthrop. gén.*, **17**, 52–65.

BELSKY, T., JOHNS, R. B., MCCARTHY, E. D., BURLINGAME, A. L., RICHTER, W., and CALVIN, M. (1965), 'Evidence of life processes in a sediment two and a half billion years old', *Nature, Lond.*, **206**, 446–447.

BERGER, R., HORNEY, A. G., and LIBBY, W. F. (1964), 'Radiocarbon dating of bone and shell from their organic components', *Science, N.Y.*, **144**, 999–1001.

BERGMANN, W. (1963), 'Geochemistry of lipids', in *Organic Geochemistry* (ed. BREGER, I. A.), Monograph No. 16 Earth Science Series, pp. 503–542. Oxford: Pergamon. Distributed by the Macmillan Co., New York.

BLUMENTALS, A., and SWAIN, F. M. (1956), 'Comparison of amino acids obtained by acid hydrolysis of lake sediments, Central Minnesota', *Bull. geol. Soc. Am.*, **67**, 1673.

BØGGILD, O. B. (1930), 'Shell structure of mollusks', *K. danske Vidensk. Selsk. Skr.*, Raekke 9, **2**, 233–326.

BOYD, W. C., and BOYD, L. G. (1937), 'Blood grouping tests on 300 mummies', *J. Immunol.*, **32**, 307–319.

BREGER, I. A., ed. (1963), *Organic Geochemistry*, Monograph No. 16 Earth Science Series. Oxford: Pergamon. Distributed by the Macmillan Co., New York.

BRENNER, H. (1939), 'Paläophysiologische Untersuchungen an der fossilen Muskulatur aus der eozänen Braunkohle des Geiseltales bei Halle (Saale)', *Nova Acta Leopoldina*, VII, 95–118.

BRICTEUX-GRÉGOIRE, S., FLORKIN, M., and GRÉGOIRE, C. (1968), 'Prism conchiolin of modern or fossil molluscan shells', *Comp. Biochem. Physiol.*, **24**, 567–572.

BRUNSKILL, G. J., and VALLENTYNE, J. R. (1966), 'Amino acids in a Silurian evaporite', *Verh. Int. Verein Limnol.*, **16**, 490–491.

BURLINGAME, A. L., and SIMONEIT, B. R. (1968a), 'Analysis of the mineral entrapped fatty acids isolated from the Green River formation', *Nature, Lond.*, **218**, 252–256.

BURLINGAME, A. L., and SIMONEIT, B. R. (1968b), 'Isoprenoid fatty acids isolated from the Kerogen matrix of the Green River formation (Eocene)', *Science, N.Y.*, **160**, 531–533.

CANDELA, P. B. (1939), 'Blood group tests on stains, mummified tissues and cancellous bone', *Am. J. phys. Anthropol.*, **25**, 187–214.

CLARKE, F. W., and WHEELER, W. C. (1922), *The Inorganic Constituents of Marine Invertebrates.* U.S. Geol. Survey, Prof. Paper No. 124.

CLARKE, R. H. (1967), 'Amino acids in recent sediments off south-east Devon, England', *Nature, Lond.*, **213**, 1003–1005.

CLOUD, P. E., jun. (1965), 'Significance of the Gunflint (Precambrian) microflora', *Science, N.Y.*, **148**, 27–35.

COOK, S. F., BROOKS, S. T., and EZRA-COHN, H. E. (1962), 'Histological studies on fossil bone', *J. Paleont.*, **36**, 483–494.

COOK, S. F., and HEIZER, R. F. (1953), 'Archaeological dating by chemical analysis of bone', *SWest. J. Anthrop.*, **9**, 231–238.

COSSLETT, V. E., and NIXON, W. C. (1960), *X-ray Microscopy.* London: Cambridge University Press.

DAS, S. K., DOBERENZ, A. R., and WYCKOFF, R. W. G. (1967), 'Lipids in fossils' *Comp. Biochem. Physiol.*, **23**, 519–525.

DEGENS, E. T., DEUSER, W. G., and HAEDRICH, R. L. (1969), 'Molecular structure and composition of fish otoliths', *Marine Biology, Int. J. on Life in Oceans and Coastal Waters*, **2**, 105–113.

DEGENS, E. T., EMERY, K. O., and REUTER, J. H. (1963), 'Organic materials in recent and ancient sediments', *Neues Jb. Geol. Paläont. Mh.*, **1963**, 231–248.

DEGENS, E. T., JOHANNESSON, B. W., and MEYER, R. W. (1967), 'Mineralization processes in molluscs and their paleontological significance', *Naturwissenschaften*, **54**, 638–640.

DEGENS, E. T., and LOVE, S. (1965), 'Comparative studies of amino acids in shell structures of *Gyraulus trochiformis* Stahl, from the tertiary of Steinheim, Germany', *Nature, Lond.*, **205**, 876–878.

DEGENS, E. T., and SPENCER, D. W. (1966), Woods Hole Oceanog. Inst., Ref. No. 66–27, Unpublished manuscript.

DEGENS, E. T., SPENCER, D. W., and PARKER, R. H. (1967), 'Paleobiochemistry of molluscan shell proteins', *Comp. Biochem. Physiol.*, **20**, 553–579.

DENNIS, R. L. (1969), 'Fossil mycelium with clamp connections from the Middle Pennsylvanian', *Science, N.Y.*, **163**, 670–671.

DOBERENZ, A. R. (1967), 'Ultrastructure of fossil dentinal collagen', *Calc. Tissue Res.*, **1**, 166–169.

DOBERENZ, A. R., and LUND, R. (1966), 'Evidence for collagen in a fossil of the Lower Jurassic', *Nature, Lond.*, **212**, 1502–1503.

DOBERENZ, A. R., and MATTER, P., III (1965), 'Nitrogen analyses of fossil bones', *Comp. Biochem. Physiol.*, **16**, 253–258.

DOBERENZ, A. R., MATTER, P., III, and WYCKOFF, R. W. G. (1966), 'Microcomposition of some fossil insects of Miocene Age', *Bull. So. Cal. Acad. Sci.*, **65**, 229–236.

DOBERENZ, A. R., MILLER, M. F., II, and WYCKOFF, R. W. G. (1969), 'Fossil enamel protein', *Calc. Tissue Res.*, **3**, 93–95.

DOBERENZ, A. R., and WYCKOFF, R. W. G. (1967a), 'Microstructure of fossil teeth', *J. Ultrastruct. Res.*, **18**, 166–175.

DOBERENZ, A. R., and WYCKOFF, R. W. G. (1967b), 'Fine structure in fossil collagen', *Proc. Natn. Acad. Sci. U.S.A.*, **57**, 539–541.

DODD, J. R. (1965), 'Environmental control of strontium and magnesium in *Mytilus*', *Geochim. cosmochim. Acta*, **29**, 385–398.

DODD, J. R. (1966), 'Diagenetic stability of temperature-sensitive skeletal properties in *Mytilus* from the Pleistocene of California', *Bull. geol. Soc. Am.*, **77**, 1213–1224.

DROZDOVA, T. V., and KOCHENOV, A. V. (1960), 'Organic matter of fossil fish', *Geokhimiya*, **1960**, 748–751.

DUGHI, R., and SIRUGUE, F. (1962), 'Distribution verticale des œufs d'oiseaux fossiles de l'Éocène de Basse-Provence', *Bull. Soc. géol. Fr.* (7), **4**, 69–78.

DUGHI, R., and SIRUGUE, F. (1964), 'Sur la structure des coquilles des œufs des Sauropsidés vivants ou fossiles; le genre *Psammornis* Andrews', *Bull. Soc. géol. Fr.* (7), **6**, 240–252.

DUGHI, R., and SIRUGUE, F. (1966), 'Sur la fossilisation des œufs de dinosaures', *C. r. hebd. Séanc. Acad. Sci., Paris*, **262**, 2330–2332.

EASTOE, J. E. (1955), 'Amino acid composition of mammalian collagen and gelatin', *Biochem. J.*, **61**, 589–600.

EASTOE, J. E. (1957), 'Amino acid composition of fish collagen and gelatin', *Biochem. J.*, **65**, 363–368.

EASTOE, J. E., and EASTOE, B. (1954), 'Organic constituents of mammalian compact bone', *Biochem. J.*, **57**, 453–459.

EASTOE, J. E., and LEACH, A. A. (1958), 'Amino acids of vertebrate collagen and gelatin', in *Recent Advances in Gelatin and Glue Research* (ed. STAINSBY, G.), pp. 173–178. New York: Academic Press.

EDINGER, T. (1929), 'Die fossilen Gehirne', *Ergebn. Anat., Berlin*, **28**, 1–249.

EHLERS, E. G., STILES, D. V., and BIRLE, J. D. (1965), 'Fossil bacteria in pyrite', *Science, N.Y.*, **148**, 1719–1721.

ENDRES, J. G. (1966), 'An apparatus for rapid preparation of fatty acid esters from lipids for gas chromatographic analysis', *J. Am. Oil Chem. Soc.*, **43**, 600.

EPSTEIN, S., and LOWENSTAM, H. A. (1953), 'Temperature-shell-growth relations of Recent and interglacial Pleistocene shoal-water biota from Bermuda', *J. Geol.*, **61**, 424–438.

ERBEN, H. K. (1969a), 'Neue Möglichkeiten in der paläontologischen Forschung', *Umschau*, **1969**, 78–80.

ERBEN, H. K. (1969b), 'Dinosaürer: pathologische Strukturen der Eischale als Letalfaktor', *Umschau*, **1969**, 552–553.

ERBEN, H. K., FLAJS, G., and SIEHL, A. (1968), 'Über die Schalenstruktur von Monoplacophoren', *Akad. Wiss. Lit. Mainz*, No. 1.

ERDMAN, J. G., MARLETT, E. M., and HANSON, W. E. (1956), 'Survival of amino acids in marine sediments', *Science, N.Y.*, **124**, 1026.

EVERTS, J. M. (1969), 'Fatty acid analysis of some fossil and recent bones and teeth', Thesis, University of Arizona.

EVERTS, J. M., DOBERENZ, A. R., and WYCKOFF, R. W. G. (1968), 'Fatty acids in fossil bones', *Comp. Biochem. Physiol.*, **26**, 955–962.

EZRA, H. C., and COOK, S. F. (1957), 'Amino acids in fossil human bone', *Science, N.Y.*, **126**, 80.

FARQUHAR, J. W., INSUL, W. I., jun., ROSEN, P., STOFFEL, W., and AHRENS, E. H., jun. (1959), 'Analysis of fatty acid mixtures', *Nutr. Rev.*, **17**, Suppl., 29–30.

FLORKIN, M. (1966), *Aspects Moléculaires de l'Adaptation et de la Phylogénie*. Paris.

FLORKIN, M., GRÉGOIRE, C., BRICTEUX-GRÉGOIRE, S., and SCHOFFENIELS, E. (1961), 'Conchiolines de nacres fossiles', *C. r. hebd. Séanc. Acad. Sci., Paris*, **252**, 440–442.

FOX, S. W., and HARADA, K. (1958), 'Thermal copolymerization of amino acids to a product resembling protein', *Science, N.Y.*, **128**, 1214.

FOX, S. W., and HARADA, K. (1960), 'The thermal copolymerization of amino acids common to protein', *J. Am. chem. Soc.*, **82**, 3745–3751.

FOX, S. W., HARADA, K., WOODS, K. R., and WINDSOR, C. R. (1963), 'Amino acid compositions of proteinoids', *Archs Biochem. Biophys.*, **102**, 439.

FOX, S. W., and NAKASHIMA, T. (1967), 'Fractionation and characterization of an amidated thermal 1:1:1-protenoid', *Biochim. biophys. Acta*, **140**, 155–167.

FOX, S. W., and YUYAMA, S. (1963), 'Abiotic production of primitive protein and formed microparticles', *Ann. N.Y. Acad. Sci.*, **108**, 487–494.

GEHRKE, C. W., ROACH, D., ZUMWALT, R. W., STALLING, D. L., and WALL, L. L. (1968), *Quantitative Gas–liquid Chromatography of Amino Acids in Proteins and Biological Substances*. Analytical Biochemistry Laboratories, Inc.

GILLESPIE, J. M. (1970), 'Mammoth hair: stability of α-keratin structure and constituent proteins', *Science, N.Y.*, **170**, 1100–1101.

GLAESSNER, M. F. (1969), 'Trace fossils from the Precambrian and the Basal Cambrian', *Lethaia*, **2**, 369–394.

GRANDJEAN, J., GRÉGOIRE, C., and LUTTS, A. (1964), 'Mineral components and remnants of organic structures in shells of fossil molluscs', *Bull. Acad. r. Belg. Cl. Sci.*, Sér. 5, **50**, 562–595.

GRÉGOIRE, C. (1957), 'Topography of organic components in mother-of-pearl', *J. biophys. biochem. Cytol.*, **3**, 797–808.

GRÉGOIRE, C. (1958a), 'Structure et topographie, étudiées au microscope électronique, des constituants organiques de la nacre chez 24 espèces (10 familles) de Gastéropodes et de Pélécypodes', *Archs int. Physiol. Biochim.*, **66**, 667–671.

GRÉGOIRE, C. (1958b), 'Essai de détection au microscope électronique des dentelles organiques dans les nacres fossiles (Ammonites, Nautilides, Gastéropodes et Pélécypodes)', *Archs int. Physiol. Biochim.*, **66**, 674–676.

GRÉGOIRE, C. (1959a), 'Conchiolin remnants in mother-of-pearl from fossil cephalopoda', *Nature, Lond.*, **184**, 1157–1158.

GRÉGOIRE, C. (1959b), 'A study on the remains of organic components in fossil mother-of-pearl', *Bull. Inst. r. Sci. nat. Belg.*, **35** (13), 1–14.

GRÉGOIRE, C. (1960), 'Further studies on structure of organic components in mother-of-pearl, especially in Pelecypods, I', *Bull. Inst. r. Sci. Nat. Belg.*, **36** (23), 1–22.

GRÉGOIRE, C. (1961), 'Structure of the conchiolin cases of the prisms in *Mytilus edulis* Linne', *J. biophys. biochem. Cytol.*, **9**, 395–400.

GRÉGOIRE, C. (1962), 'Submicroscopic structure of *Nautilus* shell', *Bull. Inst. r. Sci. Nat. Belg.*, **38** (49), 1–71.

GRÉGOIRE, C. (1966a), 'On organic remains in shells of Paleozoic and Mesozoic cephalopods (Nautiloids and Ammonoids)', *Bull. Inst. r. Sci. Nat. Belg.*, **42** (39), 1–36.

GRÉGOIRE, C. (1966b), 'Experimental diagenesis of the *Nautilus* shell', *Adv. in Organic Geochemistry*, Proc. 3rd Int. Conference (ed. HOBSON, G. D.), pp. 429–441.

GRÉGOIRE, C. (1967), 'Sur la structure des matrices organiques des coquilles de mollusques', *Biol. Rev.*, **42**, 653–688.

GRÉGOIRE, C. (1968), 'Experimental alteration of the *Nautilus* shell by factors involved in diagenesis and metamorphism, I. Thermal changes in conchiolin matrix of mother-of-pearl', *Bull. Inst. r. Sci. Nat. Belg.*, **44** (25), 1–69.

GRÉGOIRE, C., DUCHÂTEAU, G., and FLORKIN, M. (1950), 'Structure, étudiée au microscope électronique, des nacres décalcifiées de mollusques', *Archs int. Physiol.*, **58**, 117–120.

GRÉGOIRE, C., DUCHÂTEAU, G., and FLORKIN, M. (1955), 'La trame protidique des nacres et des perles', *Annls Inst. Océanogr.*, **31**, 1–36.

GRÉGOIRE, C., and TEICHERT, C. (1965), 'Conchiolin membranes in shell and cameral deposits of Pennsylvanian cephalopods, Oklahoma', *Okla. Geol. Notes*, **25**, 175–201.

GRÉGOIRE, C., and VOSS-FOUCART, M.-F. (1970), 'Proteins in shells of fossil cephalopods (Nautiloids and Ammonoids) and experimental simulation of their alterations', *Archs int. Physiol. Biochim.*, **78**, 191–203.

GUSTAVSON, K. H. (1953), 'Hydrothermal stability and intermolecular organization of collagens from mammalian and teleost skins', *Svensk kem. Tidskr.*, **65**, 70–77.

GUSTAVSON, K. H. (1956), *The Chemistry and Reactivity of Collagen.* New York: Academic Press.

HALL, A., and KENNEDY, W. J. (1967), 'Aragonite in fossils', *Proc. R. Soc.*, **168B**, 377–412.

HALLAM, A., and PRICE, N. B. (1966), 'Strontium contents of recent and fossil aragonitic cephalopod shells', *Nature, Lond.*, **212**, 25–27.

HARADA, K., and FOX, S. W. (1965), 'Characterization of thermal polymers of neutral α-amino acids with dicarboxylic amino acids or lysine', *Archs Biochem. Biophys.*, **109**, 49–56.

HARE, P. E. (1963), 'Amino acids in the proteins from aragonite and calcite in the shells of *Mytilus californianus*', *Science, N.Y.*, **139**, 216–217.

HARE, P. E. (1965), 'Amino acid artifacts in organic geochemistry', *Carnegie Inst. Wash. Year Book*, **64**, 232–235.

HARE, P. E., and ABELSON, P. H. (1964), 'Proteins in mollusk shells', *Carnegie Inst. Wash. Year Book*, **63**, 267–270.

HARE, P. E., and ABELSON, P. H. (1965), 'Amino acid composition of some calcified proteins', *Carnegie Inst. Wash. Year Book*, **64**, 223–232.

HARE, P. E., and ABELSON, P. H. (1968), 'Racemization of amino acids in fossil shells', *Carnegie Inst. Wash. Year Book*, **66**, 526–528.

HARE, P. E., and MITTERER, R. M. (1967), 'Nonprotein amino acids in fossil shells', *Carnegie Inst. Wash. Year Book*, **65**, 362–364.

HARE, P. E., and MITTERER, R. M. (1969), 'Laboratory simulation of amino acid diagenesis in fossils', *Carnegie Inst. Wash. Year Book*, **67**, 205–208.

HAYAKAWA, T., WINDSOR, C. R., and FOX, S. W. (1967), 'Copolymerization of the Leuchs anhydrides of the eighteen amino acids common to protein', *Archs Biochem. Biophys.*, **118**, 265–272.

HECHT, F. (1933), 'Der Verbleib der organischen Substanz der Tiere bei meerischer Einbettung', *Senckenbergiana, Berlin*, **XV**, 3–4, 165–249.

HEIJKENSKJÖLD, F., and MÖLLERBERG, H. (1958), 'Amino acids in anthracite', *Nature, Lond.*, **181**, 334–335.

HEIZER, R. F., and COOK, S. F. (1952), 'Fluorine and other chemical tests of some North American human and fossil bones', *Am. J. Phys. Anthrop.*, **10**, 289–303.

HELLER, W. (1965), 'Biochemie und Feinstruktur fossiler Knochen aus bituminösen Schichten', *Geol. Rdsch.*, **55**, 119–130.

HEYN, A. N. J. (1962), 'Electron microscope observations on the structure of calcite in avian egg shell', *J. appl. Phys.*, **33**, 2658–2659.

HEYN, A. N. J. (1963a), 'The crystalline structure of calcium carbonate in the avian egg shell', *J. Ultrastruct. Res.*, **8**, 176–188.

HEYN, A. N. J. (1963b), 'Calcification of avian egg shell', *Ann. N.Y. Acad. Sci.*, **109**, 246–250.

HO, TONG-YUN (1965), 'Amino acid composition of bone and tooth proteins in late Pleistocene mammals', *Proc. Natn. Acad. Sci. U.S.A.*, **54**, 26–31.

HO, TONG-YUN (1966), 'Isolation and amino acid composition of bone collagen in Pleistocene mammals', *Comp. Biochem. Physiol.*, **18**, 353–358.

HO, TONG-YUN (1967a), 'Amino acids of bone and dentine collagens in Pleistocene mammals', *Biochim. biophys. Acta*, **133**, 568–573.

HO, TONG-YUN (1967b), 'Imino acid contents of mammalian bone collagen and body temperature as a basis for estimation of body temperature of prehistoric mammals', *Comp. Biochem. Physiol.*, **22**, 113–119.

HOERING, T. C., and ABELSON, P. H. (1965), 'Fatty acids from the oxidation of kerogen', *Carnegie Inst. Wash. Year Book*, **64**, 218–223.

ISSACS, W. A., LITTLE, K., CURREY, J. D., and TARLO, L. B. H. (1963), 'Collagen and a cellulose-like substance in fossil dentine and bone', *Nature, Lond.*, **197**, 192.

JACKSON, T. A. (1967), 'Fossil actinomycetes in Middle Precambrian glacial varves', *Science, N.Y.*, **155**, 1003–1005.

JARCHO, S. (ed.) (1966), *Human Palaeopathology*. New Haven and London: Yale University Press.

JENSEN, J. A. (1966), 'Dinosaur eggs from the Upper Cretaceous North Horn Formation of Central Utah', *Brigham Young Univ. Geol. Studies*, **13**, 55–67.

JENSEN, J. A. (1970), 'Fossil eggs in the Lower Cretaceous of Utah', *Brigham Young Univ. Geol. Studies*, **17**, 51–65.

JONES, J. D., and VALLENTYNE, J. R. (1960), 'Biogeochemistry of organic matter, I. Polypeptides and amino acids in fossils and sediments in relation to geothermometry', *Geochim. cosmochim. Acta*, **21**, 1–34.

JOWSEY, J., and ORVIS, R. L. (1967), 'Comparative deposition of $^{45}Ca$, $^{65}Zn$, and $^{91}Y$ in bone', *Radiat. Res.*, **31**, 693–698.

JOWSEY, J., SISSONS, H. A., and VAUGHAN, J. (1956), 'The site of deposition of $Y^{91}$ in the bones of rabbits and dogs', *J. nucl. Energy*, **2**, 168–176.

KUMMEL, B., and RAUP, D. (eds.) (1965), *Handbook of Paleontological Technology*. San Francisco: Freeman.

KVENVOLDEN, K. A. (1967), 'Normal fatty acids in sediments', *J. Am. Oil Chem. Soc.*, **44**, 628–636.

KVENVOLDEN, K. A., PETERSON, E., and BROWN, F. S. (1970), 'Racemization of amino acids in sediments from Saanich Inlet, British Columbia', *Science, N.Y.*, **169**, 1079–1082.

LEACH, A. A. (1957), 'The amino acid composition of amphibian, reptile and avian gelatins', *Biochem. J.*, **67**, 83–87.

LEWIN, P. K. (1967), 'Paleo-electron microscopy of mummified tissue', *Nature, Lond.*, **213**, 416–417.

LITTLE, K., KELLY, M., and COURTS, A. (1962), 'Studies on bone matrix in normal and osteoporotic bone', *J. Bone Jt. Surg.*, **44B**, 503.

LOWENSTAM, H. A. (1954), 'Factors affecting the aragonite calcite ratios in carbonate-secreting marine organisms', *J. Geol.*, **62**, 284–324.

LOWENSTAM, H. A. (1964a), 'Sr/Ca ratio of skeletal aragonites from recent marine biota at Palau and from fossil gastropods', in *Isotopic and Cosmic Chemistry* (Eds. CRAIG, H., MILLER, S. L., and WASSERBURG, G. J.), pp. 114–132. Amsterdam: North Holland Publishing Co.

LOWENSTAM, H. A. (1964b), 'Coexisting calcites and aragonites from skeletal carbonates of marine organisms and their strontium and magnesium contents', in *Recent Res. Fields of Hydrosphere, Atmosphere, Nuclear Geochem.* Tokyo: Maruzen (distributor).

MARSHALL, C. G. A., MAY, J. W., and PERRET, C. J. (1964), 'Fossil microorganisms: possible presence in Precambrian shield of Western Australia', *Science, N.Y.*, **144**, 290–292.

MASSHOFF, W., and STOLPMANN, H. J. (1961), 'Licht und elektronenmikroskopische Untersuchungen an der Schalenhaut und Kalkschale des Hühnereies', *Z. Zellforsch. mikrosk. Anat.*, **55**, 818–832.

MATTER, P., III, DAVIDSON, F. D., and WYCKOFF, R. W. G. (1969), 'The composition of fossil oyster shell proteins', *Proc. Natn. Acad. Sci. U.S.A.*, **64**, 970–972.

MATTER, P., III, DAVIDSON, F. D., and WYCKOFF, R. W. G. (1970), 'The microstructure and composition of some Pliocene fossils', *Comp. Biochem. Physiol.*, **35**, 291–298.

MATTER, P., III, DAVIDSON, F. D., and WYCKOFF, R. W. G. (1971), unpublished data.

MEINSCHEIN, W. G., BARGHOORN, E. S., and SCHOPF, J. W. (1964), 'Biological remnants in a Precambrian sediment', *Science, N.Y.*, **145**, 262–263.

MILLER, M. F., II, and WYCKOFF, R. W. G. (1968), 'Proteins in dinosaur bones', *Proc. Natn. Acad. Sci. U.S.A.*, **60**, 176–178.

MOODIE, R. L. (1920), 'Thread moulds and bacteria in the Devonian', *Science*, N.S., **51**, 13–14.

MOODIE, R. L. (1923), *Paleopathology, An Introduction to the Study of Ancient Evidences of Disease*. Urbana: University of Illinois Press.

MUTVEI, H. (1970), 'Ultrastructure of the mineral and organic components of molluscan nacreous layers', *Biomineralization Res. Rpts*, **2**, 48–72.

NOLL, W. (1934), 'Geochemie des Strontiums', *Chemie Erde*, **8**, 507–600.

PARKER, R. B., and TOOTS, H. (1970), 'Minor elements in fossil bone', *Bull. geol. Soc. Am.*, **81**, 925–932.

PARTRIDGE, S. M. (1948), 'The chemistry of connective tissues, I. The state of combination of chondroitin sulphate in cartilage', *Biochem. J.*, **43**, 387–397.

PAWLICKI, R., KORBEL, A., and KUBIAK, H. (1966), 'Cells, collagen fibrils and vessels in dinosaur bone', *Nature, Lond.*, **211**, 655–657.

PEYER, B. (1968), *Comparative Odontology* (translated by ZANGERL, R.). Chicago: University of Chicago Press.

PIERCE, W. D. (1946), 'Fossil arthropods of California. 10. Exploring the minute world of the California asphalt deposits', *Bull. So. Calif. Acad. Sci.*, **45**, 113–132.

PIERCE, W. D. (1960), 'Fossil arthropods of California. 23. Silicified insects in Miocene nodules from the Calico mountains', *Bull. So. Calif. Acad. Sci.*, **59**, 40–49.

PIERCE, W. D. (1962), 'The significance of the petroliferous nodules of our desert mountains', *Bull. So. Calif. Acad. Sci.*, **61**, 7–14.

PIEZ, K. A. (1963), 'Amino acid chemistry of some calcified tissues', *Ann. N.Y. Acad. Sci.*, **109**, 256–268.

PRICE, N. B., and HALLAM, A. (1967), 'Variation of strontium content within shells of recent *Nautilus* and *Sepia*', *Nature, Lond.*, **215**, 1272–1274.

RACE, G. J., FRY, E. I., MATTHEWS, J. L., WAGNER, M. J., MARTIN, J. H., and LYNN, J. A. (1968), 'Ancient Nubian human bone: a chemical and ultrastructural characterization including collagen', *Am. J. phys. Anthrop.*, N.S., **28**, 157–162.

RANSON, G. (1952), 'Les huîtres et le calcaire. Calcaire et substratum organique chez les mollusques et quelques autres invertébrés marins', *C. r. hebd. Séanc. Acad. Sci., Paris*, **234**, 1485–1487.

RANSON, G. (1966), 'Substratum organique et matrice organique des prismes de la couche prismatique de la coquille de certains mollusques lamellibranches', *C. r. hebd. Séanc. Acad. Sci., Paris*, **262D**, 1280–1282.

REGÖLY-MÉREI, G. (1967), 'Palaeopathological examination of the skeletal finds of Naima Tolgoy and Hana', *Acta archaeol. Acad. Sci. Hungaricae*, **19**, 391–409.

RENAULT, B. (1899), 'Sur quelques microorganismes des combustibles fossiles', *Bull. Soc. Ind. minér. St. Étienne*, **13** (1899), **14** (1900).

RHODES, F. H. T., and BLOXAM, T. W. (1971), 'Phosphatic organisms in the Paleozoic and their evolutionary significance', *Proc. N. Amer. Paleont. Convention* 1969, pp. 1485–1513. Lawrence, Kans.: Allen Press.

ROCHE, J., RANSOM, G., and EYSSERIC-LAFON, M. (1951), 'Sur la composition des scléroprotéines des coquilles des mollusques (Conchiolines)', *C. r. Séanc. Soc. Biol.*, **145**, 1474–1477.

SCHIDLOWSKI, M. (1965), 'Probable life-forms from the Precambrian of the Witwatersrand system (South Africa)', *Nature, Lond.*, **205**, 895–896.

SCHMIDT, R. A. M., and SELLMANN, P. V. (1966), 'Mummified Pleistocene ostracods in Alaska', *Science, N.Y.*, **153**, 167–168.

SCHOPF, J. W., BARGHOORN, E. S., MASER, M. D., and GORDON, R. O. (1965), 'Electron microscopy of fossil bacteria two billion years old', *Science, N.Y.*, **149**, 1365–1367.

SCOTT, J. H., and SYMONS, N. B. B. (1964), *Introduction to Dental Anatomy*, 4th ed. Edinburgh and London: Livingstone.

SHACKLEFORD, J. M., and WYCKOFF, R. W. G. (1964), 'Collagen in fossil teeth and bones', *J. Ultrastruct. Res.*, **11**, 173–180.

SINEX, F. M., and FARIS, B. (1959), 'Isolation of gelatin from ancient bones', *Science, N.Y.*, **129**, 969.

STANTON, R. J., jun., and DODD, J. R. (1970), 'Paleoecologic techniques—comparison of faunal and geochemical analyses of Pliocene paleoenvironments, Kettleman Hills, California', *J. Paleont.*, **44**, 1092.

STEHLI, F. G. (1956), 'Shell mineralogy in Paleozoic invertebrates', *Science, N.Y.*, **123**, 1031–1032.

STEINMAN, G. (1967), 'Sequence generation in prebiological peptide synthesis', *Archs Biochem. Biophys.*, **119**, 76–82.

STOKES, W. L. (1964), 'Fossilized stomach contents of a Sauropod dinosaur', *Science, N.Y.*, **143**, 576–577.

STÜRMER, W. (1965), 'Röntgenaufnamen von einigen Fossilien aus dem Geologischen Institut der Universität Erlangen-Nürnberg', *Geol. Blätter NO-Bayern*, **15**, 217–223.

STÜRMER, W. (1966), 'Radiography in palaeontology', *Med. biol. Illust.*, **16**, 173–176.

STÜRMER, W. (1970), 'Soft parts of cephalopods and trilobites: Some surprising results of X-ray examinations of Devonian slates', *Science, N.Y.*, **170**, 1300–1302.

TANAKA, S., HATANO, H., and ITASAKA, O. (1960), 'Biochemical studies on pearl, IX. Amino acid composition of conchiolin in pearl and shell', *Bull. chem. Soc. Japan*, **33**, 543–545.

TARLO, L. B. H., and MERCER, J. R. (1966), 'Decalcified fossil dentine', *J. R. microsc. Soc.*, **86**, 137–140.

TASNÁDI-KUBACSKA, A. (1962), *Paläopathologie; Pathologie der vorzeitlichen Tiere.* Jena: Fischer.

TEREPKA, A. R. (1963), 'Structure and calcification in avian egg shell', *Expl. Cell Res.*, **30**, 171–182.

THIEME, F. P., OTTEN, C. M., and SUTTON, H. E. (1956), 'A blood typing of human skull fragments from the Pleistocene', *Am. J. phys. Anthrop.*, **14**, 437.

THURBER, D. L., KULP, J. L., HODGES, E., GAST, P. W., and WAMPLER, J. M. (1958), 'Common strontium content of the human skeleton', *Science, N.Y.*, **128**, 256–257.

TOOTS, H., and VOORHIES, M. R. (1965), 'Strontium in fossil bones and the reconstruction of food chains', *Science, N.Y.*, **149**, 854–855.

TOWE, K. M., and HARPER, C. W., jun. (1966), 'Pholidostrophiid brachiopods: origin of the nacreous luster', *Science, N.Y.*, **154**, 153–155.

TRAVIS, D. F., FRANÇOIS, C. J., BONAR, L. C., and GLIMCHER, M. J. (1967), 'Comparative studies of the organic matrices of invertebrate mineralized tissues', *J. Ultrastruct. Res.*, **18**, 519–550.

TRAVIS, D. F., and GONSALVES, M. (1969), 'Comparative ultrastructure and organization of the prismatic region of two bivalves and its possible relation to the chemical mechanism of boring', *Am. Zoologist*, **9**, 635–661.

TRISTRAM, G. R., and SMITH, R. H. (1963), 'Amino acid composition of some purified proteins', in *Advances in Protein Chemistry* (Eds. ANFINSON, C. B., ANSON, M. L., and EDSALL, J. T.), vol. 18, pp. 227–318. New York: Academic Press.

VAHL, J. (1971), 'The applicability of combined biocrystallographical and ultrastructural research methods in paleontology', *Proc. N. Am. Paleont. Convention* 1969, pp. 1535–1562. Lawrence, Kans.: Allen Press.

VALLENTYNE, J. R. (1963), 'Geochemistry of carbohydrates', in *Organic Geochemistry* (Ed. BREGER, I. A.), Monograph No. 16 Earth Science Series, pp. 456–502. Oxford: Pergamon. Distributed by the Macmillan Co., New York.

VALLENTYNE, J. R. (1964), 'Biogeochemistry of organic matter, II. Thermal reaction kinetics and transformation products of amino compounds', *Geochim. cosmochim. Acta*, **28**, 157–188.

VOIGT, E. (1949), 'Mikroskopische Untersuchungen an fossilen tierischen Weichteilen und ihre Bedeutung für Systematik und Paläobiologie', *Z. dt. geol. Ges.*, **101**, pp. 99–104.

VOSS-FOUCART, M.-F. (1968), 'Paléoprotéines des coquilles fossiles d'œufs de dinosauriens du Crétacé Supérieur de Provence', *Comp. Biochem. Physiol.*, **24**, 31–36.

Voss-Foucart, M.-F., and Grégoire, C. (1971), *Bull. Inst. r. Sci. Nat. Belg.*, in press.

Wada, K. (1966), 'Comparative biochemical study on the amino acid composition of conchiolins from calcitic and aragonitic layers', *Bull. Jap. Soc. scient. Fish.*, **32**, 295–303.

Walcott, C. D. (1915a), 'Evidences of primitive life', *Smithsonian Rept. for 1915*, p. 241.

Walcott, C. D. (1915b), 'Discovery of Algonkian bacteria', *Proc. Natn. Acad. Sci. U.S.A.*, **1**, 256.

Watabe, N., and Wilbur, K. M. (1961), 'Studies on shell formation, IX. An electron microscope study of crystal layer formation in the oyster', *J. biophys. Biochem. Cytol.*, **9**, 761–772.

Webby, B. D. (1970), 'Late Precambrian trace fossils from New South Wales', *Lethaia*, **3**, 79–109.

Widdowson, T. W. (1946), *Special or Dental Anatomy and Physiology and Dental Histology*, 7th ed., vol. II. London: Staples Press.

Wyckoff, R. W. G. (1964), 'Application de methodés physicochimiques à l'étude de fossiles', *Bull. Soc. fr. Minér. Cristallogr.*, **87**, 235–240.

Wyckoff, R. W. G. (1969), 'Composition de quelques protéines dinosauriennes', *C. r. hebd. Séanc. Acad. Sci., Paris*, **269**, 1489–1491.

Wyckoff, R. W. G. (1971), 'Trace elements and organic constituents in fossil bones and teeth', *Proc. N. Am. Paleont. Convention* 1969, pp. 1514–1524. Lawrence, Kans.: Allen Press.

Wyckoff, R. W. G., and Doberenz, A. R. (1965a), 'Electron microscopy of Rancho La Brea bone', *Proc. Natn. Acad. Sci. U.S.A.*, **53**, 230–233.

Wyckoff, R. W. G., and Doberenz, A. R. (1965b), 'Le collagène dans les dents pleistocènes', *J. Microscopie*, **4**, 271–274.

Wyckoff, R. W. G., and Doberenz, A. R. (1968), 'Strontium content of fossil teeth and bones', *Geochim. cosmochim. Acta*, **32**, 109–115.

Wyckoff, R. W. G., Doberenz, A. R., and McCaughey, W. F. (1965), 'The amino acid composition of proteins from desert-dried bone', *Biochim. biophys. Acta*, **107**, 389–390.

Wyckoff, R. W. G., Hoffman, V. J., and Matter, P., III (1963), 'Microradiography of fossilized teeth', *Science, N.Y.*, **140**, 78–80.

Wyckoff, R. W. G., McCaughey, W. F., and Doberenz, A. (1964), 'The amino acid composition of proteins from Pleistocene bones', *Biochim. biophys. Acta*, **93**, 374–377.

Wyckoff, R. W. G., Wagner, E., Matter, P., III, and Doberenz, A. R. (1963), 'Collagen in fossil bones', *Proc. Natn. Acad. Sci. U.S.A.*, **50**, 215–218.

# INDEX